The Chemistry of Water

The Chemistry of Water

Susan E. Kegley
University of California, Berkeley

Joy Andrews
California State University, Hayward

University Science Books
Sausalito, California

University Science Books
55D Gate 5 Road
Sausalito, CA 94965
FAX (415) 332-5393
univscibks@igc.org
www.uscibooks.com

Manuscript editor: Susan Giniger
Compositor: Printed Page Productions, Berkeley
Cover art: Leigh Anne McConnaughey
Cover design: Robert Ishi

This book is printed on acid-free paper

This work was supported by a grant from the National Science
Foundation, #USE-9156123 and the Camille and Henry Dreyfus
Foundation.

Library of Congress Cataloging-in-Publication Data

Kegley, Susan E., 1957–
    The chemistry of water / by Susan E. Kegley and Joy Andrews.
        p.   cm.
    ISBN 0-935702-44-X
    1. Water chemistry.   I. Andrews, Joy, 1956– . II. Title.
GB855.K44   1997
546'.22-dc21   97-34032
                                           CIP

Printed in the United States of America
10 9 8 7 6 5 4 3 2

This project was supported, in part by the
Division of Undergraduate Education of the
National Science Foundation
Opinions expressed are those of the authors
and not necessarily those of the Foundation

Contents

Preface

The purity of our water supplies is a pressing problem and will become increasingly more so in the coming years. This laboratory module is designed as an introduction to factors affecting water quality and the methods used to assess water quality. The chemical concepts covered include acid-base chemistry, spectroscopy, principles of sampling and quality control, the electrochemical techniques of potentiometry and ion selective electrodes, ion chromatography, and atomic absorption spectrophotometry. The module is constructed to be transferrable to a variety of institutions, providing multiple techniques for the analysis of individual water quality parameters. Some techniques require only a buret and a few chemicals, while others utilize sophisticated state-of-the-art instrumentation. In an undergraduate setting, the module is best carried out over three to five weeks, with the final week of the module used to introduce students to a spreadsheet and graphing program for data analysis.

This module has been tested for seven semesters in a first semester general chemistry course at the University of California, Berkeley in special laboratory sections focusing on environmental chemistry, as well as in the 1000-student regular general chemistry laboratory. Student response to the module has been overwhelmingly positive, with many students commenting that the knowledge gained has had a significant impact on their awareness of issues related to water quality and their ability to analyze data scientifically (see *The Chemical Educator,* 1996, *vol. 1*, http://journals.springer-ny.com/chedr).

This module would also work well in an upper level instrumental methods of analysis course. More advanced students will be able to delve deeper into the theory and practice of ion chromatography or atomic absorption spectrophotometry. The laboratory text provides background information on the chemistry of natural waters not readily found in a single source.

A detailed instructor's manual, complete with supplier information, stockroom prep lists, pre-lab lecture notes, instrument parameters, and transparency masters, is available to provide assistance in carrying out this experiment.

Acknowledgments

This work was supported by a grant from the National Science Foundation, #USE-9156123. The idea for creating the module originated from one of the author's participation in an interdisciplinary team-taught course on Environmental Science at Williams College. We are grateful to William College Professors David Dethier (Geology) and Hank Art (Biology) and Jay Thoman (Chemistry) for many helpful discussions on the interdisciplinary aspects of environmental science. We thank Professor Angelica Stacy at UC Berkeley for testing the experiments in the large General Chemistry laboratory sections and providing insightful comments on the manuscript and many good suggestions for improving the experiments.

Suggestions for procedural revisions were contributed by UC Berkeley Graduate Student Instructors Ramon Alvarez, Anna Cederstav, Kevin Cunningham, Gerd Kochendoerfer, Heather Clague, and Adam Safir. Kome Shomglin and Ben Gross tested the wet methods of analysis and drafted Appendix C. Billie Jean Lee helped with the figures. Marla Wilson of Printed Page Productions provided able assistance in designing and formatting the finished manuscript. We are grateful for the careful reviews provided by Professors David Jenkins (UC Berkeley, Civil and Environmental Engineering), Birgit Koehler (Williams College), Jack Steehler (Roanoke College), and Thomas Dunne (Reed College).

Finally, we would like to acknowledge the enthusiastic participation and feedback of the UC Berkeley and Cal State Hayward Environmental Chemistry students. Their help with the pilot testing of the module was invaluable.

Susan E. Kegley
Joy Andrews

To the Student

The environment is the world we live in, from isolated wilderness areas with cold mountain streams and clear air to inner cities crowded with people. The broad term "environment" includes the indoor world as well, and the study of the environment deals with nearly every aspect of life. The purity of the air we breathe and the water we drink, and the shelter of the atmosphere from damaging radiation are important factors contributing to the quality of life, not only for humans, but also for the many other plant and animal species that inhabit the Earth. There is nothing that brings this fact so close to home as the scarcity of necessities such as clean air and water. Headlines that announce "MASSIVE FOREST BURNS IN THE AMAZON DECIMATE BIODIVERSITY," "GLOBAL WARMING: IS IT REAL?" or "TOXIC WASTE DUMP DISCOVERED NEAR TOWN WELLS" point to some of the problems our society faces as we try to balance the growing human population with the limited resources of water, food, energy, and open space.

WHY STUDY ENVIRONMENTAL CHEMISTRY?

How does chemistry fit into all of this? Chemistry is the study of atoms and molecules and their interactions. Since every substance is composed of atoms, every substance is a chemical. The common perception that all chemicals are toxic is inaccurate and misleading. Indeed, *you* are composed of many chemical substances, including proteins, enzymes, fats, water, and bones. Soil is a mixture of chemicals. Air is a mixture of gaseous chemicals. The plastics that roller blades and CDs are made of and the Gore-tex™ fabric that makes a raincoat waterproof are also chemicals. Hazardous wastes are also composed of chemicals.

Since every substance is composed of atoms, every substance is a chemical.

So what separates the "good" chemicals from the "bad" ones, the toxic from the nontoxic? The toxicity of a substance is related to its *reactivity*, or ability to chemically combine with other molecules. Within an organism, toxic substances can react with molecules that are essential for life processes and alter their function. Sickness, cancer, or death can result. Other substances are not directly toxic to living things, but their chemical reactivity with *other* substances may cause additional problems. Chlorofluorocarbons (CFCs) are examples of such substances. These gases are nontoxic and unreactive until they reach the upper layer of the atmosphere, where they are exposed to ultraviolet (UV) light from the sun. The UV light produces reactive species that destroy ozone, a gas that shields the earth from the sun's damaging ultraviolet radiation. Thus, although CFCs are "nontoxic," their presence in the stratosphere is causing serious environmental problems. Choices to utilize substances that may cause environmental damage have been made by our society at large over the last 50 years. Although much of the damage is profit driven, it is the demands of the consumers that are ultimately the source of the problem.

The study of molecules and their reactivity is a fundamental aspect of chemistry that enables us to make predictions about which substances might be toxic or what reactions might prevent a substance from becoming an environmental hazard. *Environmental chemistry* entails studying the microscopic properties of atoms and molecules and looking at the effects these properties have on the macroscopic or observable environment. Chemistry can answer questions like "Why does ozone block out ultraviolet light?" or "Why are the oceans salty?" Knowing the details of molecular interactions through the study of chemistry makes it possible to predict and explain the phenomena that occur in the natural world.

Environmental chemistry is the study of molecules and their behavior in the environment.

What Do Environmental Chemists Do?

Chemistry plays a number of roles in environmental science. In a fundamental way, it enables us to understand natural processes in the environment. Knowledge of the physical and chemical properties of molecules allow prediction of reactivity patterns of chemical substances. This knowledge has enhanced our understanding of the atmosphere and the oceans and provided a basis for improving the quality of life for people everywhere.

Chemistry also provides answers to important questions such as "What pollutants are present?," "How much of the pollutant is present?," and "At what levels does the pollutant have adverse effects on ecosystems or human health?" There are now a vast array of modern analytical techniques that enable chemists to identify the substances that are present in a given sample and quantify the concentrations with great sensitivity. With this knowledge in hand, standards for acceptable levels of pollutants in

Chemistry answers many important questions about the sources of substances present in the environment and their effects on an ecosystem.

our food, water, and air can be set based on assessment of the risks inherent in exposure to the pollutant and the economic costs of compliance with these standards. A weighing of the costs and benefits of pollution cleanup is necessary to decide what we can afford (and are willing to afford) in the way of cleanup efforts. Thus chemistry also affects the *economics* of pollution control.

Environmental chemists have been instrumental in designing creative approaches to minimizing the impact of humans on nature. Several examples follow.

- Perhaps the biggest triumph of environmental science has been the development of the means to purify an essential renewable resource—water. Major advances in wastewater treatment make it possible for millions of people to live in a very small area and still maintain the quality of the surrounding rivers and oceans.

- In the 1950s and 1960s, foam generated from household detergents was a major problem in our rivers and streams. Chemists went to work on the formulation of better soaps and surfactants and developed detergents that biodegrade quickly, so foam is no longer a problem.

- Cleanup of hazardous waste from industrial processes often involves utilizing innovative technologies such as biodegradation, a technique that uses microbes to degrade the substance causing the problem. Using the techniques of genetic engineering, biochemists and molecular biologists have developed microbes that specifically digest a variety of pollutants.

Chemists help provide solutions to environmental problems.

- Contamination of groundwater with dry cleaning fluid, industrial solvents, herbicides, and pesticides is becoming increasingly more common. One of the methods used for remediation of this problem is "pump and treat," that is, pump the water out of the ground and treat it in some way to remove the pollutants. New chemical agents and materials that act as "traps" for specific pollutants are continually being developed by chemists while geologists devise better ways of modeling and tracking pollutant migration in the groundwater.

- Innovative chemists and process engineers have been responsible for reducing pollution by critically analyzing manufacturing processes to determine the optimal method of producing an item with the fewest by-products. Other approaches include discovering methods of separating and purifying components of a waste stream in such a way as to be able to reuse or even market them.

- On the energy front, chemists are working to design molecules that will transform solar energy into usable electrical energy. They are also designing new processes that will make coal a cleaner source of energy than it is at present. This "clean coal" technology will enable us to utilize the earth's coal supplies, yet minimize air pollution.

- Environmental chemists also work on understanding *why* natural systems behave the way they do. Laboratory studies on the chemical reactions that are responsible for destruction of the ozone layer in the stratosphere have allowed chemists to determine that chlorofluorocarbon refrigerants are the major source of the problem. At present, other chemists are working on formulation of new compounds that are good refrigerants but are nontoxic and not harmful to the ozone layer.

In short, chemistry provides many approaches to minimizing the impact of humans on natural ecosystems; however, the work of chemists is not done in a vacuum. It is dependent on interaction between people with specialized knowledge from many different disciplines.

The Interdisciplinary Nature of Environmental Science

The study of the environment is an interdisciplinary effort between specialists in biology, chemistry, engineering, geology, planning, and public health. It is only by working together that people can fully understand problems and devise solutions.

Environmental chemistry utilizes information from biology, geology, meteorology, and engineering to prevent and solve environmental problems.

For example, the study of groundwater pollution and cleanup is a task shared by chemists, geologists, and engineers. The geologists look at the topography of the site and determine where the water table is and the predicted pathway for spread of pollution from the site. The chemists analyze soil and water samples to determine the extent of the contamination. The engineers devise methods for extracting the contaminants from the water, aided by chemists who design new substances that act to remove the pollutants.

Assessment of a mercury-contaminated wetland involves chemists and biologists. The chemists analyze the plants, fish, birds, water, and sediments for trace levels of mercury. Often this information can be used to trace the source of the mercury. The biologists study the effects of high mercury levels on the plants and wildlife and look for food-chain connections between the different species.

The design of a new pesticide is a joint effort between entomologists, chemists, and toxicologists. The chemists design the molecular structure of the substance based on their knowledge of other pesticides that are effective. Determining the effectiveness of the pesticide is carried out by entomologists, and assessing the toxicity of a pesticide to humans and animals is carried out by toxicologists. For already-approved pesticides, public health officials monitor the long-term effects of the substance on humans to determine if exposure is correlated with increased incidence of cancer or other disease.

What We Hope You Will Take Away from This Module

This module is designed to expose you to a variety of challenges that environmental chemists encounter in determining the purity of water supplies. Our job will be to teach you how to obtain the necessary answers with as much accuracy as possible and to point you in a direction that will maximize your learning experience. We will teach you how to take samples, prepare samples for analysis, and use some sophisticated instrumentation that enables you to quantify trace levels of pollutants in environmental samples. But these skills are not the most important things we want you to learn from this experiment. What we would like most is to help you develop your ability to use the data you collect to assess a situation, that is, having the data in hand, to be able to step back from the immediacy of the numbers to look at the overview and answer the question "Would you drink this water?" Your job will be to keep an open mind, be unafraid of disagreeing with your classmates and instructors, develop your abilities to work both independently and in groups, and take broader views of scientific and environmental issues.

LABORATORY PROTOCOLS

Safety Considerations

Most activities have hazards associated with them. If you drive, there is always a chance that you may get in a car accident. If you ride a mountain bike on steep trails, you may end up tumbling head over heels into a rock pile. Driving to school can entail a sizeable risk of an automobile accident. However, if you observe certain precautions, the chance of such events actually occurring can be minimized. The same is true of working in the laboratory. During the course of this module, you will be working with strong acids, sources of ignition, and substances that are toxic if ingested or inhaled. For your safety, it is important that you learn the precautions necessary to handle these materials safely. Therefore, it is ESSENTIAL that all students and instructors read and follow the safety rules outlined below.

- EYE PROTECTION is provided as part of your laboratory equipment and must be worn AT ALL TIMES.

- Know the LOCATION and USE of the fire extinguisher, fire blanket, safety shower, and eyewash fountain.

- Know at least two emergency exit routes from the lab.

- When you are working with strong acids or bases, WEAR GLOVES. When you are finished, rinse the gloves before placing them in the trash can. Take special care not to drip or spill any strong acid or strong base on yourself, your clothing, or your neighbor. An unnoticed drop on the outside of a reagent bottle or on the desktop can lead to an expensive hole or a painful burn (Do not wear your best clothes to lab unless you are willing to sacrifice them to an acid hole!). If you do spill acid or base on your skin or clothing, the best remedy is to wash immediately with copious amounts of water. If clothing that is next to your skin is soaked with acid or base, the only effective way to get the corrosive material away from the skin is to remove the clothing and continue to rinse. Use the safety shower if the spill is extensive. All accidents or hazardous situations should be reported to your laboratory instructor immediately.

- When mixing concentrated acid with water, always add the acid to the water to minimize the heat generated. Never add water to concentrated acid.

Attention to detail in the laboratory will prevent accidents and minimize hazards.

- The most common accidents in the laboratory are cuts caused by broken glass. Clean up after yourself if you break something.

- Experiments in which corrosive or unpleasant gases are liberated should be carried out in the fume hood.

- Use only as much of any reagent as you need; waste is both expensive and polluting.

- NEVER PIPET **ANYTHING** BY MOUTH. Pipet bulbs for are provided for this purpose.

- You will be working with cultures of coliform bacteria. Avoid direct contact with the bacterial cultures and wash your hands thoroughly immediately after handling the plates.

- A laboratory refrigerator is not to be used for the storage of food, drink, or flammable materials.

- NO FOOD OR DRINK IN THE LABORATORY. It is always possible that someone did not clean up the bench after spilling something that you would not want to eat with your sandwich!

- The atomic absorption spectrophotometer produces very intense light. Safety glasses should protect you from any eye damage, but it is a good habit not to stare at the flame or the graphite furnace for any length of time.

- Clean up your work area before leaving the lab. Sponge down the benchtop, clean up any trash, and be sure all chemical waste is properly disposed of in the waste containers provided.

- The last thing you should do before leaving the lab every day is WASH YOUR HANDS to remove any traces of hazardous chemicals before you sit down to dinner.

Handling Nonrecyclable Chemicals

It would be a bit ironic if, in all our efforts to learn environmental chemistry, we added more pollutants to the local ecosystem. To avoid this, it is important to dispose of all nonrecyclable chemicals properly.

In general, get in the habit of NEVER putting any chemical substance down the drain or in the trash can unless your instructor specifically indicates that it is allowed. There will usually be several types of disposal containers available in the lab room. It is essential that each container be used only for its designated purpose, so read the labels carefully. Chemical compatibility and incompatibility have been taken into account in designing the chemical disposal schemes. Some of the types of disposal containers are:

Always dispose of chemical waste according to the directions provided in the laboratory manual.

- **Chemicals requiring special handling**
  Chemicals that should be collected separately include:
  Organic solvents and compounds
  Cyanides
  Toxic and/or heavy metals (Pb, Hg, Cd, Cu, Ni, Cr, Ag, etc.)
  Strong oxidants (H_2O_2, Cr(VI) compounds, perchlorates, etc.)

- **Aqueous solutions of dilute acids, bases, or nontoxic salts**
  Be careful of splashes when pouring waste acid into these containers.

- **Solids**
  Solid chemical by-products and chemical-soaked paper should be placed in solid waste containers located in the hoods, NOT IN THE TRASH CAN. Keep in mind chemical compatibility issues, which are also important for solids.

- **Broken glass**
  Do not put broken glass in the regular trash can. Custodians may hurt themselves when they empty the trash.

- **Paper and common garbage**
  Any waste material that is not contaminated with chemicals or has no sharp edges can go in the garbage cans provided. This avoids the expense of unnecessary hazardous waste disposal.

Field Trip Tips

The field trip will enable you, as the environmental chemist, to explore a site, decide where you wish to sample, and actually take the samples that you will analyze. Much information can be gained from actually visiting a site. For example, if you have a list of data and you have one point with unusually high levels of a pollutant, it might be tempting to write it off as a laboratory error. However, it might be that at that particular sampling location, there is a pipe discharging the pollutant! If you have visited the site, you would have seen that pipe and would be able to make a judgement about the source of the pollutant.

Always bring a watch, your lab manual, your lab notebook, and a pen *and* two pencils. Pencils may be the only writing implement that works in a downpour! A backpack and a camera may also be useful. Read any pertinent material ahead of time and familiarize yourself with the map of the area provided by your instructor.

What to Wear

Before leaving home, check the weather and dress to stay warm and dry during the field trip.

Dressing for field trips can be a very important aspect of the trip. If you wish to be comfortable during the trip, it is important to dress appropriately for being outside in whatever weather conditions exist. (Normally, labs will not be cancelled because of rain, wind, sleet, or snow.) Remember: There is no such thing as bad weather, just bad clothing. You should always wear comfortable shoes—tennis shoes or the equivalent. If it is cold or wet, wear *wool* or polypropylene or the equivalent. Wool and some synthetics stay warm, even when wet, whereas there is nothing quite as cold as wet cotton! If it is raining, bring a completely waterproof raincoat and an umbrella. Rain pants are also useful. Borrow these items from a friend if you do not have them. An umbrella alone is not a replacement for a good raincoat with a hood. There are many times in which you will need to have both hands free to carry out a task. If it is cold, wear a wool or synthetic sweater and bring gloves and a hat. In any event, wear comfortable clothes that you don't mind getting dirty.

A Note About Working in Groups

For all of the field trips and some of the experiments, there will be times when you will be working in groups of three or four people. There will be group decisions made and group efforts to accomplish a certain task. Be sure that it is indeed a GROUP EFFORT. If you tend to take charge and do things for other people, this is the time to be less assertive and let others have their say. If you are shy about making your opinions known, now is the time to be more outspoken and interactive. In any case, everyone should keep an eye on how the group work is distributed and do their best to share equally all tasks and decisions.

Some Special Safety Considerations for Field Labs

- Do not bring valuables on field trips.

- Wear sturdy shoes and clothing appropriate for the weather conditions (see *What to Wear*, above).

- You will mostly be working in groups in the field. Do not go off alone and get separated from your group.

- Be sure at least one person in your group has a watch. Keep track of time, and meet back at the designated location at the prearranged time.

- When you are on a boat, you MUST be wearing a lifejacket. Never stand up in a small boat or hang over the railings in a large boat or on a dock.

- Wear gloves when taking samples from biologically or chemically contaminated water or sediments.

Poison ivy and poison oak are common flora in many areas. Learn how to recognize them and stay away. A helpful tip for identifi- cation is "Leaves of three, let it be!" Unfortunately, in early spring or winter, all the leaves may be gone, but the twigs can still give you a very serious case of poison oak! It is difficult to recognize poison oak twigs, so it is best to stay out of ANY brush. Even if YOU are not allergic to poison oak, it is possible that your friends are. Be aware that once you have the oil of the plant on your skin or your clothes, you continue to spread it until you wash it off with SOAP and WATER. If you know you have contacted poison oak or poison ivy, wash the affected area as soon as possible with soap and water. If no soap is available, use any nearby source of water and scrub well.

Poison oak and poison ivy can cause a very uncomfortable rash. Learn to recognize it and steer clear!

Keeping a Laboratory Notebook

Keeping a complete lab notebook is an essential part of work in both the field and the laboratory. ALWAYS bring your lab notebook to class. In the field, you may want to make a sketch of the area you are sampling and you certainly will want to record where you sampled, using as much detail as possible so someone else can understand exactly where the sample came from. Record your data, actions, and observations permanently, in indelible ink. Date each page of your notebook.

In the laboratory, your lab notebook is a permanent record of your ideas, activities, observations, and conclusions. You should write down any changes you made in the lab procedures, any problems you encountered, dilutions made, sample numbers, calculations, and values you record from an instrumental analysis of samples. You will also put the answers to the prelab questions in the lab notebook. Organize your work and write neatly so you can find pertinent information in the future, but do not be such a neat

The laboratory notebook should contain enough infor- mation that another person could repeat the experiment from your notes.

SUBJECT	NAME
Calcium Analysis by AAS	Jennifer J.
	DATE 2/27/97

SAMPLE NOTEBOOK ENTRY

2/27/97   AA – Ca^{+2}

Using sample 9A from Lake Anza
(collected on 2/5/97)
Testing for Ca^{2+}.

- Standardize

1) Matrix Modification

Blank- 4mL DDW + 1mL KCl
Standards- 4mL standard + 1mL KCl
Sample- 4ml Sample + 1ml KCl

2) Standardize

- 5 solutions : 1ppm, 2ppm, 5ppm, 10ppm, 15ppm
- Note: only use 1-10ppm solns. for standard curve - 15ppm introduced significant curvature.
- Generated standard curve in instrument and also using absorbances:

A 422.7	Conc
.008	1 ppm
.023	2
.055	5
.099	10

- Plotted using EXCEL & stored as Sample 9A_Mon.dat. Result of linear fit:

  A = 0.012 × conc. + 0.00006
  Corr Coeff = 0.9998 ⇒ good fit, so use

3) Run Samples

- Transfer ~10ml to clean 50ml beaker
- $[Ca^{2+}]$ > 10ppm ⇒ dilute
- Dilution = 5.0 ml Sample + 45.0 ml DDW
- $[Ca^{2+}]$ = 1.02 ± 0.064 (~6% error, ok)
- Accounting for dilution

  $$[Ca^{2+}] = 10.2ppm \pm 0.64$$
  Sample 9A – Lake Anza

fanatic that you do not write *anything* down for fear of it being messy. Messy is better than nothing at all. Complete sentences are unnecessary. The most important thing is to WRITE DOWN A BRIEF DESCRIPTION OF *EVERYTHING* THAT YOU DO!

On the preceeding page is a sample notebook entry from an analysis of water samples for calcium content. It is meant to be an example of how a good notebook looks. There is no reason to copy it exactly, but use it to get a general idea of the level of detail necessary in a laboratory notebook.

Chapter 1

Water Quality

Water is the most abundant liquid on this planet and is essential to the survival of every living thing. It is uniquely suited for this role, serving as a mode of transport for the substances that sustain life and as an insulator for the inhabitants of oceans and lakes. Human bodies are made up of 65%–75% water by weight, and although individuals only need 1.5–2 liters of water (1–2 quarts) daily, they could survive only a few days if completely deprived of it.

As the population of the world continues to grow with increasing rapidity, clean water for drinking is becoming ever more scarce. Although the earth's surface is nearly 71% covered by water, most of this water is salty ocean water. The amount of fresh water available on the planet for drinking, bathing, and agriculture is quite small—less than 1% of the total (see Table 1-1).

Clean water is essential for our survival.

Despite the seemingly small percentage of fresh water available for use, estimates suggest that there is enough to support 20–40 billion people on the planet. The population of our planet is likely to be limited by scarce supplies of other resources, not water. And yet, even with the present world population of nearly 6 billion people, there are water shortages in many areas. In some of these areas, there may be plenty of water in one season of the year, but none at other times. In other locations, the available water is not fit for use because of natural or human pollution.

Most communities get their drinking water from either surface water sources or from groundwater through wells. **Groundwater** is water that has percolated through the soil and has been trapped in layers of porous rock underground. The process of percolation results in filtration and purification of the water and is one of the reasons that groundwater is often used as a source of fresh water for human use.

Table 1-1: Distribution of Water on the Earth's Surface

Source	Volume (thousands of km^3)	Percent of Total Water on Earth
Oceans	1,320,000	97.3
Icecaps and glaciers	29,200	2.14
Groundwater	8,350	0.61
Fresh water lakes	125	0.009
Saline lakes and inland seas	104	0.008
Soil moisture	67	0.005
Atmospheric water	13	0.001
Rivers	1.25	0.0001

Less than 3% of water on the planet is fresh water.

Surface water is fresh water that remains on the surface of the land and does not percolate through the soil. Rivers, streams, reservoirs, and lakes are all surface waters. Surface water is replenished by rainfall and is a major source of fresh water for human use.

NATURAL CONSTITUENTS OF WATER

Water found on the earth's surface contains a plethora of dissolved substances. The predominant source of these substances is natural weathering of rocks. Weathering results when water flowing over the surface of the earth dissolves whatever minerals are present and carries the dissolved ions to the oceans, where they eventually are incorporated into sediments. Human contributions include soluble substances from mining or industrial activities or from wastewater treatment plants. Minor sources of ionic substances in natural waters include the rainout of atmospheric particles containing nitric acid (HNO_3) and sulfuric acid (H_2SO_4) from the emissions of power plants, automobiles, and other combustion-related activities, sodium chloride (NaCl) from ocean spray, and a variety of trace amounts of other substances, including dust particles made up of aluminosilicates, $Al_xSi_yO_z$, where x, y, and z are variable, depending on the mineral composition of the dust. The major natural constituents of fresh water, their relative abundances, and their impacts on water quality are shown in Appendix A.

REGULATION OF WATER QUALITY

Water Quality Standards

When you turn on the tap to get a glass of water to drink, do you think about the purity of the water that comes out of the tap? It is quite likely that you assume that your water is clean enough to not pose a hazard to you. But how can you *know* that? And what exactly is *safe*?

There are many naturally occurring substances found in water. Some of these substances are not harmful; in fact, some constituents such as calcium, magnesium, iron, and iodide are necessary nutrients and drinking water can even be a good source of these minerals. Other minerals such as iron, manganese, and sulfate, although not toxic, can give water a bad taste. High levels of sodium in combination with chloride will give water a salty taste. Calcium and magnesium react with carbonate, bicarbonate, sulfate, and soap to form precipitates that interfere with certain water uses.

Water naturally contains many dissolved minerals.

Of the natural minerals frequently found in water supplies, fluoride, nitrate, and sodium are perhaps the most important contaminants in terms of negative effects on human health. Fluoride in quantities greater than 6 mg/L can mottle children's teeth, and at higher levels is quite toxic. Levels of sodium greater than 20 mg/L may be harmful for people on a low sodium diet. Excess nitrate (>10 mg/L) in drinking water is a serious health concern because it causes a condition called **methemoglobinemia** or "blue baby syndrome" in infants, where nitrate reacts with the blood oxygen carrier **hemoglobin** to reduce its oxygen-carrying capacity. The hemoglobin can no longer carry out its function as an oxygen transport system, with the result that infants turn blue from lack of oxygen in their blood. Because nitrate is used as a fertilizer, there is significant concern over nitrate leaching into water supplies.

The Safe Drinking Water Act

The Safe Drinking Water Act of 1974 was an effort to protect public drinking water supplies in the U.S. The act was amended in 1986 and renamed the Public Health Service Act (PHSA). The law requires the Environmental Protection Agency (EPA) Administrator to "publish maximum contaminant level goals and promulgate national primary drinking water regulations for each contaminant which, in the judgment of the Administrator, may have any adverse effect on the health of persons." The EPA sets goals for the **maximum contaminant levels (MCLs)** for both **primary** and **secondary contaminants**. Primary contaminants are substances that may pose serious health hazards in drinking water. Secondary contaminants affect mainly the aesthetic

The Safe Drinking Water Act establishes maximum allowable levels of contaminants in a drinking water supply.

Table 1-2: National Primary Drinking Water Standards for Inorganic and Microbiological Constituents of Water

Contaminant	Maximum Contaminant Level (MCL)
Inorganic	
Antimony (Sb)	0.006 mg/L
Asbestos	7 MFL (million fibers per liter longer than 10 μm)
Arsenic (As)	0.05 mg/L
Barium (Ba)	2 mg/L
Beryllium (Be)	0.004 mg/L
Cadmium (Cd)	0.005 mg/L
Chromium (Cr)	0.1 mg/L
Cyanide (CN^-)	0.2 mg/L
Fluoride (F^-)	4 mg/L (secondary MCL of 2 mg/L triggers public notice)
Mercury (Hg)	0.002 mg/L
Nickel (Ni)	0.1 mg/L
Nitrate (NO_3^-, as N)	10 mg/L
Nitrite (NO_2^-, as N)	1 mg/L
Total Nitrite/Nitrate	10 mg/L
Selenium (Se)	0.05 mg/L
Sulfate (SO_4^{-2})	Regulation deferred because of relatively low health risks
Thallium (Tl)	0.002 mg/L
Lead (Pb)	0.015 mg/L (No MCL set. This number represents the Action Level (AL), the level at which the authorities must do something to remove the contaminant.)
Copper (Cu)	1.3 mg/L (No MCL set. This is the AL.)
Microbiological	
Total coliform	≤1 per 100 mL
Fecal coliform	0 per 100 mL
Giardia lamblia	0 per 100 mL
Viruses	0 per 100 mL

Source: United States Environmental Protection Agency.

qualities of drinking water such as appearance, taste, and smell (see Table 1-2 and Table 1-3). The law requires regular monitoring of water quality and provides enforcement procedures for violations.

Table 1-3: National Secondary Drinking Water Standards

Contaminant	Maximum Contaminant Level (MCL)
Chloride (Cl^-)	250 mg/L
Fluoride (F^-)	2.0 mg/L
Sodium (Na^+)	20 mg/L (health advisory)
Surfactants (soaps and detergents)	0.5 mg/L
Iron (Fe)	0.3 mg/L
Manganese (Mn)	0.05 mg/L
pH	6.5–8.5 pH units
Sulfate (SO_4^{-2})	250 mg/L
Total dissolved solids (TDS)	500 mg/L
Zinc (Zn)	5.0 mg/L
Color	15 color units
Corrosivity	Noncorrosive
Odor	3 (threshold odor number)

Source: United States Environmental Protection Agency.

The Public Health Service Act requires the EPA to set standards for MCLs allowable in drinking water. If the water delivered from the water treatment plant does not meet these standards on a regular basis, the source water is usually too polluted to be used for a public drinking water supply. How are the standards set? In the case of the primary standards, the MCL of a particular contaminant is determined by assessing the risk to the public from the presence of the contaminant and determining the highest level at which ill effects from the substance are small. This process involves studying the health effects of the pollutant on humans (if these effects are known) and carrying out animal studies to determine the potential for toxicity.

YOUR MISSION

In this experiment, your mission will be to determine if a local water source would be drinkable, or **potable**, and if not, what substances would need to be removed to make it so. In order to answer this question, you will first visit the site to observe potential sources of contamination. Your class will then design a sampling plan that will provide information about water quality in different locations at the water source. On return to the laboratory, you will test your samples for the presence of some of the major ionic components of water, compare the data to the Environmental Protection Agency's Water Quality Standards for drinking water, and write up your results in a form that could be used to make a recommendation about the water source to the local Water Board or the government agency that regulates the use of water supplies in your area.

The experiments contained in this laboratory manual provide instructions for the analysis of the following water quality parameters:

- Total dissolved solids

- Dissolved oxygen

- Coliform bacteria

- pH and alkalinity

- Common anions such as fluoride, chloride, nitrate, sulfate, and phosphate

- Common cations such as sodium, calcium, and magnesium

Learning Goals

This module is designed to introduce you to the chemistry of natural waters, the analytical methods used to evaluate water quality, and the techniques of data analysis. Specifically, by the end of the module, you should have learned the following:

- How to design and implement a sampling plan

- The chemical principles behind the methods used to assess water quality and how to apply them in analyzing and interpreting data

- How to critically evaluate and interpret water quality data

- How to write a comprehensive report that effectively conveys the data to the outside world

Chapter 2

Sampling

The first step in any environmental analysis is to design a sampling plan that will provide answers to the question you are asking about the site. In this module, the question of interest is "Would you drink this water?" If the answer is "No," it is important to ask "What substances would have to be removed from the source before it is drinkable?" Collection and analysis of samples should provide representative data about the water source and its inflows. It is also important to make observations at the site that might provide clues to possible sources of pollutants such as human or animal activity, inflows to the water source, erosion, etc.

Once the sampling plan has been decided, it is necessary to carry it out with extreme care. Dirty equipment or poor sampling techniques can completely ruin the most carefully laid sampling plans!

A sampling plan is designed to provide information about general water quality, as well as potential sources of pollutants.

SAMPLING CONSIDERATIONS

Designing a Sampling Plan

A good sampling plan will provide a snapshot of the water source on the day the samples were taken. The sampling plan should be objective and give a balanced view of the site, while providing some detailed information on the water quality of inflows to the source. In order to best determine the quality of a water source, samples should be taken under a variety of conditions over a long time period.

Three basic sampling approaches are typically used in taking environmental samples: **judgemental** sampling, **systematic** sampling, and **random** sampling (see Figure 2-1).

Judgemental sampling is a result of a bias of the sampler and usually occurs when one samples where the concentration of pollutants is thought to be high (or low). This type of sampling is not usually representative of the entire site but is useful in that it can give a "worst case" or "best case" scenario of a pollutant source.

Systematic sampling usually involves dividing the site into equal sized areas and sampling each area. Creating a measured grid or regular pattern of sample sites is an easy way to set up a systematic sampling scheme.

Random sampling involves selecting sample sites with no particular pattern or reason. The choice of sites is truly a random process. Combinations of random sampling with judgemental or systematic sampling can also be carried out. For example, a site might be divided into four large areas with samples taken randomly within these areas.

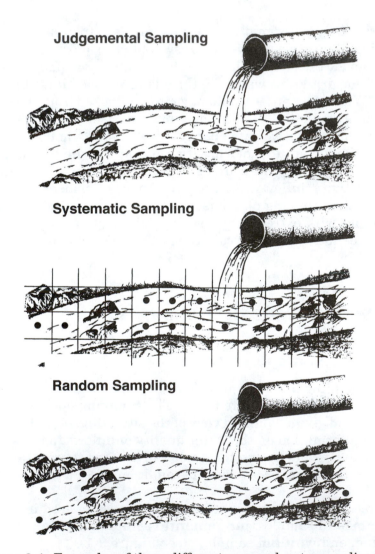

Figure 2-1: Examples of three different approaches to sampling are shown above. The solid dots represent sample sites. (Reprinted with permission from Lawrence H. Keith, *Environmental Sampling and Analysis*, © Lewis Publishers, Inc., Chelsea, MI 1991.)

When the number of samples that can realistically be taken is limited by time or cost, **composite** sampling, where equal quantities of sample taken from different sites are combined and analyzed as a single sample, is often used. For example, a lake might be divided into sections and a 1-L composite sample consisting of 250 mL from four different sites would be analyzed. A river might be sampled at several different times of the day to collect a composite over time. Composite sampling is a fast and cost-effective way to survey a site. Samples that contain high levels of pollutants point the investigation toward the problem areas which can then be sampled more carefully.

Composite sampling makes it possible to sample a larger area with fewer samples.

Sample Replicates

Murphy's Law of sampling says that the number of samples that can realistically be analyzed is always less than the number needed to fully characterize the site. It is a fact of environmental sampling that there are rarely enough resources (time and money) to analyze every corner of the site. Adding to the problem is the need to take multiple samples, called **replicates**, at most, if not all sites. Replicate samples are very important for two reasons:

- Contaminated equipment or poor lab technique can result in unexpectedly high or low results. The results could also be *real*, but there is no way to know for sure unless a replicate sample gives the same unexpected result.

- Sample bottles may break or leak. In many cases, it may not be possible to sample the same site with precisely the same water chemistry, and the information is lost forever.

In the design of any sampling plan, it is necessary to make hard choices about where to sample and how many samples to take. Although replicate samples may not be taken at *every* sample site, it is *critical* to take them at important locations such as inflows and outflows. Careful planning can maximize the data collected and result in a data set that does indeed provide a good profile of the water quality at the site.

The careful design of a sampling plan is an important first step in assessment of the potability of a drinking water supply.

Deciding Where to Sample

At a specific site, there are many factors to consider when designing a sampling plan, including, but not limited to, the following:

- What is the topography of the land? A map of the area is always useful to determine what potential pollutant sources are uphill from the water source. If groundwater is being sampled, a map of the water table will be very helpful if it is available.

- What are the effects of dilution? If an inflow is very contaminated, but consists of only a trickle of contaminated water, there may not be a problem.

- What are potential sources of the pollution? The history of the site is useful information to have if it is available.

- If a pollutant source is present, might the pollutant have migrated due to natural forces?

- What are the effects of normal weather patterns, for example, currents, prevailing winds, stream flows, etc.?

- What are the effects of recent weather, for example, rain, snow, high stream flows, seasonal temperature changes, etc.?

- How many replicate samples at each site should be taken?

- Should composite samples or individual samples be taken?

- How many samples should be taken?

Sample Blanks

Blanks allow the analyst to ascertain that the samples taken are not contaminated from other sources.

In order to be sure samples taken are truly representative of the site, and that contamination is not causing artificially high or low readings, it is normal to include some *blank* samples in the analysis. Blank samples help to determine if there are extraneous sources of pollutants. For example, if the sample bottles are not completely clean, some contamination might be introduced into the sample. If any of the reagents or glassware being used in the procedure contain trace amounts of the pollutant being tested for, it is important to know this and correct for it. There are several different types of blanks, described in detail below.

- **Field Blank**
  A field blank is taken when there might be the possibility of air contamination (including rain or fog) of the sample during the sampling process. For water sampling, the field blank is a bottle of pure deionized water that is taken to the sample site and left open and exposed to the air for the same amount of time that the real samples are exposed to the air. The bottle is then taken back to the lab and analyzed in the same manner as the real samples.

- **Trip Blank**
  The trip blank is used to measure possible contamination from the container or any preserving reagents added to the samples during sample handling and storage. For water sampling, the trip blank is a bottle of pure deionized water that is taken to the sample site but never opened, returned to the lab, and analyzed with the other samples. The trip blank would show possible leaching of substances from the sample container.

- **Equipment Blank**

  The equipment blank gives an indication of possible contamination from sampling equipment and is taken by rinsing the equipment after it has been cleaned and analyzing the rinse solution for possible contamination.

- **Background Samples**

  When sampling an area suspected to be contaminated by human activities, it is important to know what the *natural background* concentration of the pollutant is in that area. For example, there is concern over mercury pollution in South San Francisco Bay. Certain industries are suspect as the polluters, but it is also true that mercury occurs naturally in the area and was mined there until 1975. Determination of background mercury levels in this case is important to assess the amount of damage caused by industry. A background sample is taken *near* the site of interest to determine if the site is truly contaminated by human activities or is just normal for the area.

FIELD WORK: PRELAB ASSIGNMENT

Reading

Laboratory Protocols: pages 5–9

Water Quality: pages 13–18

Sampling: pages 19–28

Field Measurements: pages 29–45

Coliform bacteria: pages 48–51

1   Where are the eye wash and safety shower located in your laboratory room?

2   If you spill acid on yourself while washing glassware, what should you do?

3   Using the map of your site provided by your instructor, select 10 sample sites. Make an "X" on the map at each site.

4   For each site selected, state the information you expect to gain by sampling this site and the importance of sampling there.

5   When you do the test for coliform bacteria, you will need to return to the lab to count the bacterial colonies. When must this be done in order to obtain a reliable answer?

6   What important safety precaution should be taken after counting the bacterial colonies?

LABORATORY PROCEDURES: SAMPLING

Preparation of Glassware and Sample Bottles

The best containers for liquid water samples are clean, sterile (if bacterial analyses will be done) plastic bottles. The cleaning procedure below will serve to sterilize the containers. If you are analyzing snow or ice, heavy-duty Ziploc® bags that have not previously been used are good sample containers. Transfer the snow sample to a clean plastic bottle once it has melted.

It is possible to introduce contamination into your sample from the sampling utensils, sample container, or the lab glassware used to process the sample. Thus if you wish to obtain reliable data, it is necessary to have scrupulously clean laboratory glassware and sampling containers. A good general procedure for cleaning equipment for water analysis is given as follows. Glassware cleaning is best done the week before the sampling trip.

Acid Washing the Sample Container, Sampling Utensils, and Labware

In preparation for collecting water samples and carrying out the coliform bacteria analysis, clean the following glassware:

- One 500-mL plastic sample bottle for taking the sample on the field trip
- One 500-mL filter flask
- 10-, 50-, and 100-mL graduated cylinders
- One 100-mL volumetric flask
- Two plastic sample bottles, 125 mL and 250 mL sizes

> **⚠ Caution** The acids used for cleaning are concentrated and will cause burns if spilled on the skin or in the eyes. It is imperative that goggles be worn at all times. If you spill acid on your skin, flush the affected area with copious amounts of water. Know where the closest eyewash, shower, and sink are located. If you spill acid on the lab bench, put a *small* amount of baking soda on the spill to neutralize the acid and then wipe up the excess baking soda. DO NOT leave unlabeled containers of acid on your laboratory benchtop.

1   Inspect the glassware. Does it look dirty? Is it stained with rust or something unidentifiable? Do you know its history, i.e., what was in there last? If the item is stained, go to Step 2. If the item looks relatively clean, go to Step 3. If you are in doubt, ask your instructor. If you *know* that *you* washed it earlier, you may not need to wash it again. If someone else *told* you that he/she washed it, wash it again. The rule is: Do not trust anyone else to wash your glassware! You need to *know* that it is clean.

2   If the item is very dirty, it might be necessary to dip it in a No-Chromix™ bath, a mixture of concentrated sulfuric acid and ammonium persulfate.

DANGER!! This reagent is very corrosive. Wear GOGGLES (not just safety glasses) and gloves when using this reagent.

Caution

Any work with No-Chromix™ should be done with the assistance of your instructor. If you spill any on yourself, flush the area immediately with water. After dipping the item in the cleaning bath, rinse the glassware first in a bucket of rinse water, then take it to the sink for the remainder of the washing process. You may skip to Step 7 if you have washed your glassware with No-Chromix™ first.

3   For glassware that you can fit a brush into, scrub it with soap and water with a bottle brush. If your brush is old and looks dirty, get a new one from your instructor. Rinse with tap water.

4   Obtain 25 mL of 8 M nitric acid (HNO_3), and **carefully** carry it back to your lab station. Rinse the item with the acid, being sure to wet all inside surfaces. This serves to oxidize and solubilize any oxidizable materials. You do not have to fill the item to capacity with the wash or rinse solution. Between 5 and 25 mL should be enough. If the glassware is not visibly dirty, you may reuse the acid wash solution on other items.

5   Rinse with tap water twice. DO NOT REUSE THE RINSE WATER!

6   Obtain 25 mL of 1.2 M hydrochloric acid (HCl), and **carefully** carry it back to your lab station. Rinse the item with the acid, again being sure to wet all inside surfaces. You do not have to fill the item to capacity with the wash or rinse solution. Between 5 and 25 mL should be enough. If the glassware is not visibly dirty, you may reuse the acid wash solution on other items.

7   Rinse with tap water at least three times. DO NOT REUSE THE RINSE WATER!

8   Do a final rinse with deionized water, at least three times. DO NOT REUSE THE RINSE WATER! Do not dry items with a towel, as this will contaminate them again. Do not touch the part of the item that will come in contact with the sample with your fingers, as they are quite "dirty" if you are doing analysis for trace levels of contaminants. Store the items upside down so no dust or airborne bacteria fall onto the surfaces that might come in contact with a sample.

Do not dump acids down the drain. Dispose of all used acid solutions in the waste containers provided.

Waste
Disposal

Sampling the Water Supply

Observations

The first goal when you arrive at the site is to familiarize yourself with it. This is best done by dividing into groups of four to five students and walking around the site. Write down your observations in your laboratory notebook. This may be your only chance to visit the site, so write down as thorough a description as you can. At a minimum, note the following:

1 Take particular note of potential pollution sources such as animals, birds, human activities, pipes that might indicate a pollutant source, debris, and the behavior of any aquatic life.

2 Note the condition of the water. Is it clear? Cloudy? Smelly? Foamy?

3 If you do not already have a map of the site, sketch one in your notebook.

4 Find the actual locations of the sample sites you selected for the prelab assignment and determine whether it is feasible or useful to sample at these locations. Discuss the site with your group members and decide among yourselves which locations would be best to sample.

5 Write brief descriptions of these sites in your notebook to bring back to the group as a whole for discussion.

Group Discussion to Determine Final Sampling Plan

Each group should report their findings to the entire class. If your group missed a useful observation, be sure to write it down in your notebook. The group as a whole should then decide the following:

- Number of samples to be taken

- Locations of sample sites

- Number of replicates for each sample

Each group will then be assigned to either take samples or gather field measurements (see pages 37 and 45).

Labeling the Sample Bottles

The bottles used for sampling may already be labeled with a numbering scheme designed by your instructor. In order to avoid confusion during sample analysis in the laboratory, careful labeling is *extremely* important. Some do's and don'ts follow:

- **DON'T** create your own labeling scheme if one already exists. Use the number on the sample bottle and refer to it in your notebook.

- **DO** write a brief description of the site on the bottle, using compass points and landmarks to describe the site. For example, the description "Left of first tree from path" leaves quite a bit to be desired in terms of precision. A better description would be "North of downed tree on east side of lake" or "Pool downstream of waterfall near bridge." Check with classmates in other groups to see if they can understand what your descriptive labels mean.

- **DON'T** use water-soluble ink to write on the sample bottles. Use an indelible marking pen instead (provided by your instructor). There is nothing more frustrating than to return to the lab and find that the descriptions you so carefully wrote on the bottles have dissolved!

Taking the Sample: A Few Tips

- **VERY IMPORTANT!** Label the sample bottle with *indelible* ink *before* immersing it in the water. Pens do not write well on wet surfaces, and water-soluble inks will disappear instantly in water.

- Remember that your fingers are "dirty" in the sense that they will add ionic substances to the water sample. Do not touch the inside of the sample container or the container lid.

- When taking a water sample, position the bottle so the mouth of the bottle is pointing upstream and your fingers (holding the bottle) are downstream of the bottle opening. Fill the bottle to the top and cap it underwater if possible.

- When taking a water sample, first rinse the sample bottle with a bit of the sample. This is a good habit to get into for both sampling and carrying out the analytical procedures.

Preserving and Storing the Sample

It might not always be possible to analyze a sample as soon as you arrive back in the laboratory and for this module, you will likely keep your sample for several weeks. It is important to store the sample correctly, so no changes occur in the sample. The presence of oxygen (if the sample bottle is not completely full) and bacteria in a sample are the biggest culprits for changing the concentrations of different analytes. Filtration of the sample through 0.45 μm pore-size filters will remove most bacteria and slow the degradation of the sample. See page 49 for filtration instructions. For water samples, observe the following guidelines:

- Storage before filtration should be at 4°C for no more than 2 days.

- Storage after filtration should be at 4°C for no more than 30 days.

Checklist for the Field Trip

By the end of the field trip you should have done the following:

1 Written down general observations about the site.

2 Taken water samples in carefully labeled bottles.

3 Written down brief descriptions of the sites from which your group collected water samples.

4 Taken any field measurements desired (dissolved oxygen, conductivity), carefully recording site descriptions and the measurements obtained in your laboratory notebook.

Chapter 3

Field Measurements

With the advent of portable instruments and test kits capable of
providing accurate measurements of water quality parameters,
many people are now measuring certain parameters in the field,
often eliminating the necessity of sampling altogether. The most
important advantage of these field techniques is ease of use.
Additionally, transport of samples to the laboratory can often
result in a change in concentration of an analyte, particularly
when an analyte is gaseous and can easily be removed or added to
the solution through agitation or changes in temperature. This
section provides background information on field techniques for
measuring dissolved oxygen and total dissolved solids.

DISSOLVED OXYGEN (DO)

A parameter related to water quality that can be measured easily,
in the field is the amount of **dissolved oxygen** (abbreviated **DO**)
in the water. Although it is not directly related to drinking water
quality, the availability of oxygen in surface waters is an important
factor in determining what aquatic life can survive and is an
indicator of the concentration of nutrients and organic matter in
the water. Low levels of dissolved oxygen frequently indicate a high
concentration of decaying organic matter in the water. As bacteria
digest organic matter, they use up oxygen, leaving little for the
other aquatic creatures.

*The level of dissolved oxygen
in a water supply is indicative
of the concentration of nutrients and organic matter in the
water.*

Factors Affecting the Solubility
of Oxygen in Water

The amount of any gas dissolved in water is related to the pressure
of the gas above the water. At sea level, the pressure of the air is
1 atmosphere. With the air composed of approximately 78%

nitrogen and 21% oxygen, the maximum amount of oxygen pressure over a body of water is only 0.21 atmosphere.

Oxygen is only slightly soluble in water. When an aqueous solution is saturated with air at 1 atmosphere of pressure and 25°C, the amount of oxygen present in the solution is only 8.32 mg/L. The amount of dissolved oxygen in a body of water must be above a certain level to sustain life, with most fish requiring at least 5–6 mg/L of DO for survival.

Oxygen is only slightly soluble in water.

The amount of oxygen dissolved in water is limited both by inherent physical solubility and by the fact that waters are frequently not at maximum saturation. There are three factors that affect the inherent solubility of oxygen in water:

1 **Temperature**   As temperature increases, the amount of oxygen (or any gas) dissolved in water decreases. For example, you probably know that a hot soft drink will lose its fizz (caused by dissolved carbon dioxide gas) much more quickly than a cold one. If you heat water on a stove, air bubbles appear long before the water boils, as the dissolved gases escape. Figure 3-1 shows the relationship between temperature and solubility of O_2.

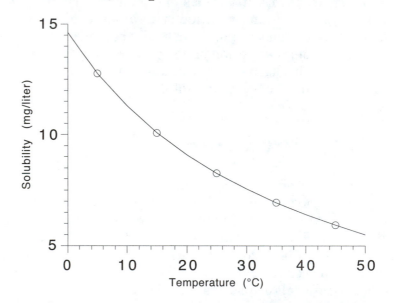

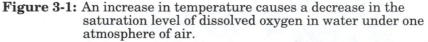

Figure 3-1: An increase in temperature causes a decrease in the saturation level of dissolved oxygen in water under one atmosphere of air.

The inverse relationship of the solubility of oxygen to temperature is responsible for the concern over **thermal pollution**. Many industries use water as a coolant for processes that generate heat. Thermoelectric power generation is a particularly significant problem in this regard, as many millions of cubic meters of warm water annually are pumped back into rivers and streams by the power generation industry. Urban areas also contribute to thermal pollution when rain

High water temperatures reduce the amount of dissolved oxygen present in solution.

falls and runs off of paved surfaces that have been heated by the sun. Similar heating occurs when rivers or streams are channelized and the paved river bottoms transfer heat to the water. Removal of trees from river and stream banks can also reduce shading and increase water temperature. In the summer when the temperature of surface water is already high because of sunlight and high air temperatures, the added thermal pollution from runoff and industry can be disastrous for fish, insect larvae, and other aquatic life. The problem is not only that the levels of DO are lower at higher water temperatures. A compounding factor is that at higher temperatures, the metabolic rate of aquatic life increases, thus increasing the demand for DO when the supply is lowest.

2 **Partial Pressure of Oxygen in Contact with Water**   The concentration of oxygen in water is directly related to the **partial pressure** of oxygen in the air that is in contact with the water. The partial pressure of a gas is the fraction of one component (e.g., oxygen) present in a gas mixture (e.g., the atmosphere). Although this is a relatively constant value at sea level, a change to a higher elevation can significantly decrease the amount of available dissolved oxygen. As elevation increases, the pressure exerted by the atmosphere on the surface of a body of water is less because there are fewer gas molecules and therefore fewer collisions of gas molecules with the surface of a lake or stream. The atmospheric pressure decreases steadily with increasing altitude. The pressure change can be estimated using the empirical relationship:

At high altitudes, less oxygen is dissolved in water because the partial pressure of oxygen in the atmosphere is low.

$$P = P_0 - [(1.15 \times 10^{-4} \text{ atm/m}) \times H]$$

where P is the atmospheric pressure (in atmospheres) at altitude H (in meters), and P_0 is the atmospheric pressure at sea level (in atm). The net result is lower levels of dissolved oxygen at higher altitude.

3 **Salinity of the Water**   The concentration of salts in water, or **salinity**, affects how much gas dissolves in water. Typically, the solubility of gases (O_2, CO_2) *decreases* with increasing salinity. This phenomenon plays a major role in regulating how much O_2 and CO_2 are taken up by the oceans.

The three factors mentioned above relate to the physical limits of gas solubility in water. However, even in cold water at sea level, it is still possible that the maximum concentration of DO has not been reached. There are three other factors that control the *availability* of oxygen in natural waters:

1 **Amount of Mixing Between Air and Water**   The mixing of air and water is brought about by exposure of the surface of water to oxygen in the air. Anything that increases this exposure will increase the amount of dissolved oxygen present,

Water turbulence enhances the rate of dissolution of oxygen in water.

with turbulence playing the most important role in increasing the amount of DO in a body of water. In rivers and streams, rapids and waterfalls serve as an effective way to mix oxygen and water (see Figure 3-2). In lakes and the oceans, wind-blown waves create surface turbulence that mixes oxygen into the water. Even though deep water is colder than surface water and should theoretically be able to hold more dissolved oxygen, surface water will nearly always contain more DO because of increased turbulence. Ice-covered water is usually oxygen deficient because little exchange takes place between air and water. Artificial aeration is often used to increase dissolved oxygen content of wastewater during the treatment process.

Figure 3-2: Rapids provide a natural means for maximizing the amount of oxygen dissolved in water.

Waters with high biochemical oxygen demand have little oxygen available for fish and other aquatic animals.

2   Biochemical Oxygen Demand (BOD)   The biochemical oxygen demand of a water sample is a measurement of the concentration of biodegradable organic matter in the water. The utilization of organic matter in water by microorganisms requires oxygen for the biochemical metabolic processes that occur and when the concentration of organic matter is high, the dissolved oxygen in the water is used up by rapidly growing microorganisms, leaving little oxygen for use by the fish and other aquatic organisms such as the zooplankton that form the basis of the food chain in natural waters.

$$\text{organic matter (C, H, O)} + O_2 \longrightarrow CO_2 + H_2O$$

An example of where this is problematic is the algal blooms that occur when a stream, river, or estuary has been contaminated with runoff containing high levels of organic matter and nutrients such as nitrate and phosphate. Fertilizer and feedlot runoff and sewage wastes are primary causes of this problem. When the algae die, the bacteria that degrade the algae use up all available oxygen.

3 **Chemical Oxygen Demand (COD)**   The chemical oxygen demand of a water sample is a measure of the total concentration of organic matter in the water. Many organic substances such as animal wastes and compost are easily decomposed by chemical reactions with oxygen in water, a process that uses up precious dissolved oxygen. The presence of high levels of organic contaminants in water can lead to decreased levels of DO. A situation where a high COD is placed on a body of water is when runoff from feedlots enters a lake or river. This nutrient-rich mixture is also rich in organic compounds that are easily degraded to CO_2 and H_2O by reaction with oxygen. The oxygen used up in the reaction is no longer available for use by aquatic organisms and the level of DO decreases.

There are also processes that can increase the level of DO *above* the saturation concentration for a particular temperature. The most important of these is photosynthesis by algae. During periods of rapid algal growth, oxygen is produced so quickly that it cannot be dissipated fast enough to maintain saturation levels of oxygen in the water. The result is a body of water that is **supersaturated** with oxygen, that is, the concentration of dissolved oxygen is *above* the equilibrium value. Water in which an algal bloom is in full swing will often appear to "fizz" as dissolved oxygen escapes from the water on a sunny day.

Water can also be supersaturated with oxygen, such as when algae in the water are growing rapidly, producing oxygen from photosynthesis.

MEASUREMENT OF DISSOLVED OXYGEN

The oxidizing properties of oxygen can be used to determine its concentration in water. Two field techniques for measuring dissolved oxygen will be discussed, one using an oxygen-sensing electrode and the other using the Winkler titration.

The Oxygen-Sensing Membrane Electrode

The dissolved oxygen meter (Figure 3-3) consists of an electrode that is shielded from the sample solution by a membrane that is selectively permeable to oxygen. When oxygen diffuses through the membrane and reaches the electrode, a transfer of electrons takes place between the electrode and the oxygen molecule.

The reaction at the cathode is

$$O_2(aq) + 4 H^+ + 4e^- \longrightarrow H_2O(l)$$

where oxygen reacts with hydrogen ion supplied by a buffer solution contained within the membrane. The reaction at the anode is

$$2 Pb^0(s) \longrightarrow 4e^- + 2 Pb^{+2}(aq)$$

This transfer of electrons causes a current to flow through a current-measuring device consisting of an ammeter and a reference electrode, where the amount of current is proportional to the amount of dissolved oxygen in the sample. The electrode is calibrated by using solutions (or moist air) of known oxygen concentration. Concentrations of dissolved oxygen are reported in mg/L, with saturation at 8.32 mg/L at 25°C. Levels of DO below 5 mg/L are marginal for the survival of fish and other aquatic life.

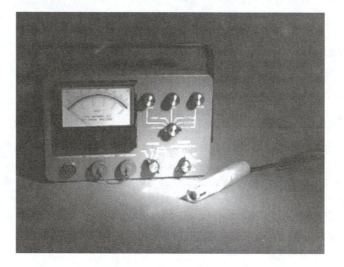

Figure 3-3: The dissolved oxygen meter permits precise field measurements of dissolved oxygen in water.

The Winkler Titration

In the Winkler titration, there is a 1:4 mole ratio of dissolved oxygen in the sample to thiosulfate used in the titration.

Concentrations of dissolved oxygen can also be measured titrimetrically, using a method called the Winkler titration. The Winkler titration requires some laboratory equipment and more time to complete than the membrane electrode method.

The method, originally developed by Winkler in 1888[1], is an indirect method that relies on a series of redox reactions initiated by the oxygen dissolved in the solution. In the initial step of the

---

1.   L. W. Winkler, *Berlin Deutche Chemische Geschichte*, 1888, v. 21, p. 2843.

reaction, oxygen reacts chemically with Mn^{+2} to remove two electrons and form the oxidized species Mn^{+4}, which precipitates out as insoluble MnO_2 under basic conditions.

$$Mn^{+2}(aq) + 2\,OH^-(aq) + 1/2\,O_2\,(g) \longrightarrow MnO_2(s) + H_2O\,(l)$$

The solution is then treated with sulfuric acid to dissolve the MnO_2. Sodium iodide is added and is oxidized by Mn^{+4} to form iodine (I_2).

$$MnO_2(s) + 4H^+(aq) + 2I^-(aq) \longrightarrow I_2\,(aq) + Mn^{+2}\,(l) + 2H_2O\,(l)$$

It is the iodine whose concentration is finally measured using a titrimetric method. A starch indicator is added, which turns blue in the presence of iodine. The now blue/purple solution is titrated with a known concentration of sodium thiosulfate solution until the blue color disappears, signaling the transformation of all I_2 to I^-.

$$I_2\,(aq) + 2\,S_2O_3^{-2}\,(aq) \longrightarrow S_4O_6^{-2}\,(aq) + 2\,I^-\,(aq)$$

The amount of oxygen present in the sample initially can be calculated from the stoichiometry of the reactions which dictates that for every mole of O_2 present, 2 moles of I_2 are produced, which requires 4 moles of thiosulfate to titrate to the endpoint. Using this 1:4 ratio of O_2 to $S_2O_3^{-2}$, the volume and concentration of the thiosulfate solution used to titrate to the endpoint can be used to determine the amount of O_2 initially present. A sample calculation follows.

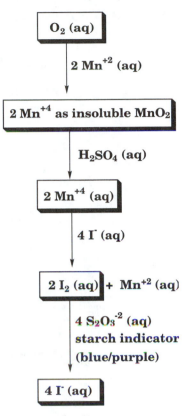

Example Calculation

A 50.00 mL water sample taken from the outfall of a sewage treatment plant required 0.47 mL of a 0.040 M $Na_2S_2O_3$ solution to reach the endpoint of the titration. How much DO was initially present in the water sample in mg/L?

The first step is to convert mL of thiosulfate solution to mmoles.

$$0.47\ \text{mL} \times \frac{0.040\ \text{mmol}\ S_2O_3^{-2}}{\text{mL}} = 0.0188\ \text{mmol}\ S_2O_3^{-2}$$

From the stoichiometry of the reaction, we know that for every mole of O_2 initially present, 4 moles of thiosulfate are used up in the titration. To obtain the number of moles of O_2 initially present, the moles of $S_2O_3^{-2}$ must be divided by 4.

$$0.0188\ \text{mmol}\ S_2O_3^{-2} \times \frac{1\ \text{mmol}\ O_2}{4\ \text{mmol}\ S_2O_3^{-2}} = 0.0047\ \text{mmol}\ O_2$$

The moles of O_2 are then converted to mg of O_2 and divided by the volume of sample analyzed to give the concentration of dissolved oxygen in mg/L (ppm).

$$0.0047 \text{ mmol } O_2 \times \frac{32 \text{ mg}}{\text{mmol } O_2} \times \frac{1}{0.05000 \text{ L}} = \frac{3.01 \text{ mg } O_2}{L} = 3.01 \text{ ppm}$$

The advantage of the Winkler method of DO determination is that it is less expensive to use a few chemicals than to purchase a dissolved oxygen meter. The disadvantages include difficulty of field use and reactions of interfering substances with the Winkler reagents. If samples are collected in the field and brought back to the laboratory for testing, care must be taken to avoid increasing or decreasing the DO in the sample by additional exposure to air or agitation of the solution. The membrane electrode method is less prone to problems with interferences and is more convenient for field work.

LABORATORY PROCEDURES: DISSOLVED OXYGEN

Dissolved Oxygen (DO), Method 1: Oxygen-Sensing Electrode

Objective: To measure the amount of dissolved oxygen in water using an oxygen-sensing electrode. See pages 29–33 for more information on dissolved oxygen and page 33 for the theory behind the oxygen-sensing electrode.

Calibrating the Meter

The following instructions are representative of the steps necessary to calibrate a meter. Actual instructions for a specific meter will vary. Refer to the manual for your instrument.

1   Turn the meter on and allow it to stabilize for 10–15 minutes.

2   Zero the reading using the mechanical zero control. With some meters, this step is unnecessary.

3   Measure the ambient temperature and, using Table 3-1, determine the amount of oxygen contained in the air at the ambient temperature. This is called the **saturation value**.

4   Using Table 3-2, correct the saturation value for altitude if the location is significantly above sea level (more than 500 ft). For example, if the temperature is 22°C, the saturation value from Table 3-1 is found to be 8.74 mg/L. If the location being sampled is at 2300 ft. elevation, this saturation value must be multiplied times ~0.92 to obtain the corrected value of 8.04 mg/L.

5   Calibrate the electrode for oxygen saturation by placing it in a small bottle with a wet paper towel in the bottom. This creates an atmosphere of 100% humidity around the electrode, similar to the conditions when the electrode is submerged in water. Turn the meter to the read position, allow it to stabilize, and adjust the reading to the corrected saturation value calculated in step 4.

6   **DO NOT** turn the meter off after calibration or it will lose the calibration.

Measurement of DO at Sample Sites

Immerse the DO probe in the water at the site to be tested. If there is a stirring motor on your instrument, be sure that it is on. Allow the meter to stabilize, then record the reading in your laboratory notebook, next to a description of the site. The depth at which the reading is taken has a significant effect on the outcome of the measurement. (Why?) In order for measurements to be comparable, all readings should be taken at approximately the same depth.

Think about your readings as you take them. Do they make sense? Typical readings of surface-level water will be between 5-12 mg/L, depending on temperature. If your readings are significantly out of this range, there may be a problem with the meter, or you may have discovered something interesting!

Table 3-1: Saturation Values of Dissolved Oxygen as a Function of Temperature

Temp. (°C)	DO (mg/L)	Temp. (°C)	DO (mg/L)	Temp. (°C)	DO (mg/L)
0	14.62	16	9.87	32	7.31
1	14.22	17	9.67	33	7.18
2	13.83	18	9.47	34	7.07
3	13.46	19	9.28	35	6.95
4	13.11	20	9.09	36	6.84
5	12.77	21	8.92	37	6.73
6	12.45	22	8.74	38	6.62
7	12.14	23	8.58	39	6.52
8	11.84	24	8.42	40	6.41
9	11.56	25	8.26	41	6.31
10	11.29	26	8.11	42	6.21
11	11.03	27	7.97	43	6.12
12	10.78	28	7.83	44	6.02
13	10.54	29	7.69	45	5.93
14	10.31	30	7.56	46	5.84
15	10.08	31	7.43	47	5.74

Table 3-2: Altitude Correction for Dissolved Oxygen Measurements

Altitude (ft)	Altitude (m)	Correction Factor[a]	Altitude (ft)	Altitude (m)	Correction Factor[a]
-276	-84	1.01	5067	1544	0.83
0	0	1.00	5391	1643	0.82
278	85	0.99	5717	1743	0.81
558	170	0.98	6047	1843	0.80
841	256	0.97	6381	1945	0.79
1126	343	0.96	6717	2047	0.78
1413	431	0.95	7058	2151	0.77
1703	519	0.94	7401	2256	0.76
1995	608	0.93	7749	2362	0.75
2290	698	0.92	8100	2469	0.74
2587	789	0.91	8455	2577	0.73
2887	880	0.90	8815	2687	0.72
3190	972	0.89	9178	2797	0.71
3496	1066	0.88	9545	2909	0.70
3804	1160	0.87	9917	3023	0.69
4115	1254	0.86	10293	3137	0.68
4430	1350	0.85	10673	3253	0.67
4747	1447	0.84	11058	3371	0.66

[a]   Multiply the saturation value at sea level (from Table 3-1) times this correction factor to obtain the saturation value at the specified altitude.

Dissolved Oxygen (DO), Method 2: Winkler Titration[2,3]

Objective: To determine the amount of dissolved oxygen in water using a titrimetric method. See pages 29–33 for more information on dissolved oxygen and page 34 for the theory behind the Winkler titration.

Note

Be careful to not introduce extra oxygen into the sample by shaking it or exposing it to air. Samples should be collected in glass stoppered bottles with no air gap and analyzed as soon as possible after collection. During the sample analysis, you should work quickly and minimize the time the solution is exposed to air.

1   Take 250–300 mL of sample in a collection bottle and use a pipet to add 1 mL of $MnSO_4$ solution and then 1 mL of the alkali–iodide–azide reagent.

Note

Do not put the pipet into the water sample! Contamination of the sample will result.

2   Stopper the mixture, being sure to exclude air bubbles, and invert a few times (~15–30) to mix. A manganese hydroxide precipitate will form. Allow the precipitate to settle until it is halfway down in the bottle, then shake again.

3   After the precipitate again settles to half way down the bottle, add 2 mL of concentrated H_2SO_4, re-stopper and mix by inversions until all precipitate dissolves.

Caution

Concentrated sulfuric acid (H_2SO_4) causes severe burns if spilled on the skin or in the eyes. It is imperative that goggles be worn at all times. If you spill acid on your skin, flush the affected area with copious amounts of water. Know where the closest eyewash, shower, and sink are located. If you spill acid on the lab bench, put a *small* amount of baking soda on the spill to neutralize the acid and then wipe up the excess baking soda. DO NOT leave unlabeled beakers of acid on your laboratory benchtop. If you spill acid on yourself, flush the affected area with copious amounts of water.

4   Measure 203 mL of the treated sample (or enough to correspond to 200 mL of the original sample plus a proportionate amount of added reagents) with a graduated cylinder, and gently pour it into a 500-mL Erlenmeyer flask.

2. D. Jenkins, V. L. Snoeyink, J. F. Ferguson, and J. O. Leckie, *Water Chemistry Laboratory Manual*, 3rd ed. (John Wiley and Sons, New York, 1980).

3. A. D. Eaton, L. S. Clesceri, and A. E. Greenberg, eds. *Standard Methods for the Examination of Water and Wastewater*, 19th ed. (American Public Health Association, Washington, DC, 1995), Method 4500-O.

5   Titrate with 0.100 M sodium thiosulfate ($Na_2S_2O_3$) solution until the solution turns a pale straw color.

6   Add 1 mL of starch solution and continue the titration until the blue color disappears at the endpoint.

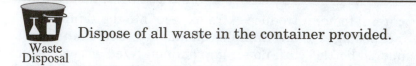 Dispose of all waste in the container provided.

Waste
Disposal

Calculations

Using the volume of sodium thiosulfate solution required to titrate to the endpoint, calculate back to the concentration of oxygen in the sample in mg/L. See page 35 for more information on calculations related to the Winkler titration.

TOTAL DISSOLVED SOLIDS (TDS)

Introduction

The total amount of dissolved chemical species in water is called **total dissolved solids**, abbreviated **TDS**, and is a good general measure of the concentration of ionic substances in water. In general, fresh water has less than 1,500 mg/L of TDS, brackish water between 1500 and 5000 mg/L TDS, and saline water above 5000 mg/L. Seawater has a TDS content of 30,000–40,000 mg/L. Table 3-3 lists typical concentrations of the major ionic constituents of an average river. For seawater, where the concentrations of ions are very high, the amount of dissolved solids in solution is expressed as **salinity**, or milligrams of solids per gram of solution. The oceans and some inland lakes or seas such as the Great Salt Lake in Utah, Mono Lake in California, or the Dead Sea between Israel and Jordan have a large quantity of dissolved solids (mostly sodium chloride or calcium carbonate) and thus have high salinity. Table 3-4 lists typical concentrations of the major constituents of seawater.

The amount of total dissolved solids is an indicator of the concentration of ionic substances present in a water supply.

Table 3-3: Composition of Average River Water

Constituent	Concentration (mg/L)
Bicarbonate (HCO_3^-)	58
Calcium (Ca^{+2})	15
Sulfate (SO_4^{-2})	11
Chloride (Cl^-)	8
Sodium (Na^+)	4
Magnesium (Mg^{+2})	4
Potassium (K^+)	2

Source: R. O. Gill, *Chemical Fundamentals of Geology*, (Unwin Hyman Ltd., London, 1989), p. 92.

Fresh water in rivers, streams, and rainwater typically has very low salinity. **Estuarine waters**, where rivers and streams mix with ocean water, have intermediate salinity that decreases from the mouth of the river inland. Although most estuarine species can tolerate some variation in salinity, they are best adapted to a particular zone of salinity. When the salinity of a body of water increases significantly over a short period of time, many species die. Thus, it is important to ensure minimum levels of **instream flow** to maintain the ecological health of the plants and animals that live in or adjacent to a river or stream. This is particularly critical in **delta** areas where a river empties into the

Many species thrive in the intermediate salinity of a river delta zone.

ocean, since these areas are important breeding grounds for many aquatic species.

Table 3-4: Major Constituents of Seawater

Constituent	Concentration (mg/L)
Chloride (Cl^-)	18,980
Sodium (Na^+)	10,560
Sulfate (SO_4^{-2})	2,560
Magnesium (Mg^{+2})	1,272
Calcium (Ca^{+2})	400
Potassium (K^+)	380
Bicarbonate (HCO_3^-)	142
Bromide (Br^-)	65
Strontium (Sr^{+2})	13
Boron, as borate (BO_4^-)	4.6
Fluoride (F^-)	1.4
Rubidium (Rb^+)	0.2
Aluminum (Al^{+3})	0.17
Lithium (Li^+)	0.1
Barium (Ba^{+2})	0.05
Iodide (I^-)	0.05

Source: F. Van der Leeden, F. L. Troise, and D. K. Todd, *The Water Encyclopedia*, 2nd ed. (Lewis Publishers, Chelsea, MI, 1990), p. 238.

Conductimetric Determination of Total Dissolved Solids (TDS)

The amount of current that can flow through a solution is proportional to the concentration of dissolved ionic species in solution.

It is possible to obtain a field measurement of the total concentration of just the *ionic* components in solution by taking advantage of the fact that ions are charged species. A solution containing charged species will act as a conductor and permit the flow of electricity through the solution. The amount of current that flows is proportional to the concentration and types of dissolved ions in solution.

A solution of pure water has a very high resistance to the flow of electricity because the only ions that are present are trace quantities (1.0×10^{-7} M at 25°C) of H_3O^+ and OH^- arising from the **autoionization** of water.

$$2\ H_2O \xrightleftharpoons{\ \ K_{eq}\ \ } H_3O^+ \ + \ HO^-$$

The electrical **resistance** of the water is lowered by dissolution of an ionic compound in the solution. The ions provide a pathway for the flow of electrons through the solution. The **conductance** of a solution is simply the reciprocal of the resistance and is a measure of the solution's ability to conduct electricity. The unit of measure for conductance is the **Siemen (S)**. Conductances in freshwater systems are usually on the order of microsiemens (μS) (i.e., 10^{-6} S).

In the laboratory, the instrument that is used to measure total dissolved solids is a **conductivity meter**, a sensitive electrical circuit called a Wheatstone bridge that measures the conductance of the solution between two parallel plates submerged in the sample. **Conductivity** is the quantity that is usually recorded as a measure of total dissolved solids, where conductivity is *conductance per unit distance between the two parallel plates*, usually expressed as Siemens per centimeter (S/cm) or, for smaller conductivities, microsiemens per centimeter (μS/cm). Very pure deionized water has a typical conductivity of less than 1 μS/cm, whereas rainwater can have conductivity in the range of 20–40 μS/cm from dissolved ions such as nitrate and sulfate from acid rain formed from combustion emission sources. Unpolluted surface waters vary in conductivity depending on the underlying geology of the area, but are generally in the range of 30–400 μS/cm. The effluent from a wastewater treatment plant may have a high conductivity (300–1000 μS/cm) because of the ionic substances found in treated wastewater (Cl^-, Na^+, K^+, PO_4^{-3}, NO_3^-, SO_4^{-2}).

The relationship of conductivity in μS/cm to total dissolved solids in mg/L depends on the identity of the ions in the solution. For example, a solution containing 500 mg of $NaHCO_3$ per liter has a conductivity of 550 μS/cm, whereas a solution containing 500 mg of $CaSO_4$ per liter has a conductivity of 580 μS/cm. A rough guide to the conversion between μS/cm and mg/L for natural waters is that conductivity in microsiemens per centimeter is about 110–115% of the dissolved solids in mg/L.

Conductivity in μS/cm is equal to approximately 110-115% of the concentration of dissolved solids in mg/L.

Gravimetric Determination of Total Dissolved Solids (TDS)

Another method used to determine the total dissolved solids is to take a known volume of the sample of water, filter it, and carefully evaporate the water. When all the water has evaporated, a dry residue will remain consisting of the constituents that were previously dissolved in the water. The dry residue can be weighed in order to determine the weight of dissolved solids in mg per liter of water. This method is more time-intensive than a conductivity measurement, but the equipment used is less expensive.

Total dissolved solids can also be determined by evaporating a known quantity of water and weighing the remaining solids.

Salinity: A Special Case of TDS

The measurement of salinity is a special case of measuring total dissolved solids in solutions with very high salt content, such as brackish water or seawater. Salinity is a dimensionless number that expresses the mass of dissolved salts in a given mass of solution.

$$\text{Salinity} = \frac{\text{mass of dissolved salts (mg)}}{\text{mass of solution (g)}}$$

The ratio of milligrams per gram is also called **parts per thousand**, abbreviated **ppt.**

The only truly accurate method of determining salinity is the gravimetric method of weighing the solution, evaporating the water, and reweighing to determine the amount of dissolved salt present. This is a rather time- and labor-intensive process and can only be carried out in a laboratory. Fortunately, the true salinity of a solution can be correlated with its conductivity. A simple measurement with a conductivity meter can then replace the process of weighing, evaporating, and reweighing. The salinity scale on a commercial conductivity meter usually ranges from 0–40 ppt. Seawater has a typical salinity of 35 ppt.

LABORATORY PROCEDURES: TOTAL DISSOLVED SOLIDS (TDS)

Method 1: Measuring TDS Using Conductivity

Objective: To determine the total concentration of dissolved ionic species in a sample by measuring the conductivity of the solution. Conductivity is a measure of how well electricity is conducted by the solution, a property affected by the concentration of the ions present in solution as well as the temperature. Although this test is not specific for a particular ionic substance, it gives an indication of the total concentration of dissolved ions in solution. For more information on total dissolved solids, see page 42.

Measuring Conductivity at Sample Sites

Immerse the conductivity probe in the water at the site to be tested. Allow the meter to stabilize, then record the reading in your laboratory notebook, next to a description of the site.

Think about your readings as you take them. Do they make sense? Typical values for the conductivity of surface waters can be found on page 42. If your readings are significantly out of this range, there may be a problem with the meter, or you may have discovered something interesting!

Measuring Conductivity in the Laboratory

It is standard procedure to measure conductivity using an *unfiltered* sample.

Note

All glassware used in this experiment should be acid-washed and well rinsed according to the procedures on page 25.

Note

1   Rinse the conductivity cell with deionized water and shake it gently to remove the wash water. Place the cell into the sample, making sure that it is covered completely with solution. Agitate it gently to remove air bubbles. Measure the temperature of the solution with a thermometer and record the temperature in your laboratory notebook.

2   Turn the meter on. If your meter is of the Wheatstone bridge type, balance the cell to obtain a null reading on the meter. On other meters, the value can simply be read from the display. In your notebook, record the value for the conductivity of the solution in microsiemens per centimeter (μS/cm).

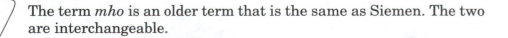

The term *mho* is an older term that is the same as Siemen. The two are interchangeable.

Note

3   If the temperature of the solution is not 25°C, correct the reading using the equation below.

$$C_{25} = C_t (1 - 0.025 \, \Delta t)$$

where C_{25} is the conductivity at 25°C, C_t is the conductivity measured at the sample temperature, and Δt is the temperature difference between the sample and 25°. If the temperature is less than 25°C, Δt is negative. If the temperature is greater than 25°C, Δt is positive.

4   Turn the meter off. Thoroughly rinse the cell with deionized water and place it in a beaker of clean deionized water.

Waste
Disposal

If you dispose of your sample at this point, you have not added any chemicals, so it may be disposed of down the drain.

Method 2: Measuring TDS Using a Gravimetric Method

Objective: To determine the concentration of total dissolved solids in a solution by filtering the sample, evaporating the water at 103–105°C, and weighing the residue remaining after evaporation of the water. The concentration of total dissolved solids determined by this method is given in milligrams per liter and includes not only ionic substances, but also non-ionic organic substances that are water soluble. For more information on total dissolved solids, see page 42.

1 Label a clean 150-mL beaker with your sample number and place it in a drying oven at 103–105°C for at least 1 hour. Remove the beaker from the oven, place it in a desiccator until cool, then weigh the beaker to the nearest 0.1 mg, recording the weight in your laboratory notebook.

2 Filter 50.0 mL of sample through 0.45 μm filter paper into the clean, dry beaker.

3 Place the beaker on wire gauze over a Bunsen burner and heat the sample to just below boiling to reduce the volume of the sample to approximately 10 mL. Do not allow the sample to boil over or splatter.

4 Allow the beaker to cool and place it in the oven at 103–105°C overnight or until the next laboratory period.

5 Remove the beaker from the oven and place it in a desiccator to cool to room temperature, then reweigh to the nearest 0.1 mg, recording the weight in your laboratory notebook.

Calculations

Calculate the total dissolved solids in milligrams of solids per liter of solution.

COLIFORM BACTERIA

An important indicator of water quality is the number of bacteria present in the water. Although it would be difficult to determine the presence of *all* bacteria in a sample, certain types of microorganisms can serve as indicators of pollution. Coliform bacteria are defined as "comprising all aerobic and facultative anaerobic, Gram-negative, non-spore-forming, rod-shaped bacteria that produce a dark colony with a metallic sheen within 24 hours at 35 °C on an Endo-type medium containing lactose."[4] This means that these rod-shaped bacteria prefer to live in the presence of oxygen (aerobic), but are capable of living in the absence of oxygen (facultative anaerobic) by utilizing different metabolic pathways. The term "gram-negative" refers to a staining method used to identify bacteria with a certain type of cell wall under a microscope. "Non-spore-forming" means that these bacteria cannot form spores that are resistant to desiccation and excess heat or cold. Coliform bacteria are ubiquitous, living in the soil, in sediments, and in the intestines of animals. The **total coliform** test is specific for all types of coliform bacteria, and is a good indicator of general water purity.

Fecal coliform bacteria are indicators of pollution of a water supply by fecal material from warm-blooded animals.

Fecal coliform bacteria are a subset of the total coliforms and represent only those that live in the intestines of warm-blooded animals such as humans, dogs, deer, raccoons, skunks, cows, etc. If there are any fecal coliforms in a water sample, it has likely been contaminated by sewage or animal wastes. Although fecal coliform bacteria might not make you sick, their presence indicates that more harmful bacteria and viruses that accompany fecal material might also be present.

For the purpose of determining the purity of drinking water, swimming holes, treated effluent from sewage treatment plants, etc., the **membrane filter technique** is widely used to analyze for the presence of coliform bacteria. A known volume of sample is filtered through a filter that is capable of trapping all bacteria. Individual bacterial cells will grow on the filter into visible colonies over a 24-hour period. When the appropriate food source containing lactose and an indicator dye is used, colonies of coliform bacteria appear greenish-black with a metallic sheen for the *total coliform* test and dark blue in the *fecal coliform* test. The color arises from the interaction of a metabolite of lactose that reacts with the dye that is in the culture medium.

Each individual bacterial cell grows into a visible colony on this fecal coliform test plate. For this sample, 100 mL of water was filtered, with a resulting coliform count of 3/100 mL.

4. L. S. Clesceri, A. E. Greenberg, and R. R. Trussell, eds., *Standard Methods for the Examination of Water and Wastewater*, 17th ed. (American Public Health Association, 1989), p. 9-66.

LABORATORY PROCEDURES: COLIFORM BACTERIA

Fecal Coliform Bacteria

Objective: To determine if fecal coliform bacteria are present in the water supply using the membrane filter technique. For information on coliform bacteria, see page 48.

1   Assemble the filter apparatus consisting of a *clean* 500 mL filter flask and a magnetic filter funnel. Each person should use his/her own filter flask, acid washed according to the procedures on page 25. The magnetic funnels might be shared equipment. If you are not the first person to filter a sample, be sure the previous person has left you with a *clean* filter funnel!

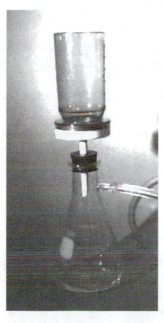

2   Obtain a petri plate and nutrient broth for *fecal coliform* (the broth is blue). Label the plate with the sample number, your initials, and "Fec" for fecal coliform. Open the sterile packet containing a very thin filter membrane and a thick nutrient pad. Using forceps, **carefully (!)** place the sterile filter membrane into the filtering apparatus. The membranes are quite fragile and very expensive, so be careful with this step. Place the thick nutrient pad into the petri plate.

3   Crack open the vial of nutrient broth by wrapping it in a paper towel and popping it open between your fingers. Pour the broth onto the pad in the petri plate, tapping it lightly on the surface to drain all the broth.

4   Shake the sample well and measure 100 mL into a *clean* 100 mL graduated cylinder, then filter it through the prepared filter membrane, using the hand pump to apply suction. If the sample contains particles of algae or sediment, allow it to settle before filtration.

5   Using forceps, remove the membrane filter carefully from the apparatus and place it on the nutrient pad in the petri plate, being careful not to trap any air bubbles between the membrane and the pad.

6 Save the filtered sample that remains in the bottom of the filter flask in a *clean* plastic bottle. **LABEL** the bottle with the sample number followed by an "F" to indicate that it has been filtered. Place this sample in the refrigerator or give it to your instructor.

7 Label the petri plate and place it in the 44.5° incubator labeled "Fecal Coliform" for 24 hours ±2 hours.

8 After incubating the plate for 22–26 hours, open the plate and count the number of dark blue bacterial colonies on the plate. **COUNT ONLY THE BLUE ONES, NO OTHERS**. Each colony represents a single bacterial cell that has grown into a colony, thus the dots are representative of the number of bacteria in 100 mL of sample. Report your fecal coliform counts as number of bacteria per 100 mL. Be sure to record the data on the sheet taped to the front of the incubator, as well as in your notebook.

Waste Disposal   After counting the colonies, discard your (open) petri plate in the bucket of chlorine bleach solution provided. **WASH YOUR HANDS**.

Note   Do not forget to return to the lab the next day (24 hours ±2 hours) to count the colonies on the plate.

Total Coliform Bacteria

Objective: To determine if coliform bacteria are present in the water supply using the membrane filter technique. For information on coliform bacteria, see page 48.

1 Obtain a petri plate and nutrient broth for *total coliform* (the broth is in the dark-brown ampoules and is pink when you pour it out). Label the plate with the sample number, your initials, and "Tot" for total coliform. Open the sterile packet containing a filter membrane and a nutrient pad. Using forceps, **carefully (!)** place the sterile filter membrane into the filtering apparatus. You do not need to clean the filter flask in between your two samples, but you should rinse the filter funnel with deionized water. Place the nutrient pad into the petri plate as before.

2 Crack open a vial of nutrient broth by wrapping it in a paper towel and popping it open between your fingers. Pour the broth onto the pad in the petri plate, tapping it lightly on the surface to drain all the broth.

3 Shake the sample well and measure 10 mL into a 100-mL graduated cylinder, dilute it with deionized water up to 100 mL, then filter it through the prepared filter membrane, using the hand pump to apply suction. There is no need to save the filtrate from the total coliform test.

4 Remove the membrane filter carefully from the apparatus and place it on the nutrient pad in the petri plate, being careful not to trap any air bubbles between the membrane and the pad.

5 Carefully rinse the filter funnel with copious amounts of deionized water to leave it clean for the next person. Clean the glassware you used so you know it is clean for the next time you need it.

6 Label the petri plate and place it in the 35° incubator labeled "Total Coliform" for 24 hours ±2 hours.

7 After incubating the plate for 22–26 hours, open the plate and count the number of green metallic bacterial colonies on the plate. **COUNT ONLY THE RED ONES WITH THE GREEN METALLIC SHEEN, NO OTHERS.** Each colony represents a single bacterial cell that has grown into a colony, thus the dots are representative of the number of bacteria in 10 mL of sample. Multiply by 10 to report your fecal coliform counts as number of bacteria per 100 mL. Be sure to record the data on the sheet taped to the front of the incubator, as well as in your notebook. If there are too many to count, report it as TNC (Too Numerous to Count).

Waste Disposal — After counting the colonies, discard your (open) petri plate in the bucket of chlorine bleach solution provided. **WASH YOUR HANDS**.

Note — Do not forget to return to the lab the next day (24 hours ±2 hours) to count the colonies on the plate.

Chapter 4

Acids and Bases

Did you ever wonder what makes your mouth pucker up when you taste a lime? Or why soap tastes bitter and feels slippery? These characteristics of common household substances are associated with their acidity or basicity. Acids and bases have a profound impact on drinking water quality.

THE ROLE OF pH IN WATER QUALITY

Exactly what *are* acids and bases? The **Brønsted-Lowry** definition characterizes an **acid** as a substance that can act as a proton (H^+) donor. In the equation below, hydrochloric acid, HCl, is acting as an acid, donating a proton to water.

Acid: $$HCl + H_2O \longrightarrow H_3O^+ + Cl^-$$

A **base** is a substance that can act as a proton acceptor. In the following equation, aqueous ammonia is acting as a base, accepting a proton from water. Note that water can act as either an acid or a base.

Base: $$NH_3 + H_2O \longrightarrow NH_4^+ + HO^-$$

An acidic solution contains an excess of **hydronium ions (H_3O^+)** over the number of **hydroxide ions (OH^-)** and a basic solution contains an excess of hydroxide ions over hydronium ions.

Acids that are commonly encountered around the home include substances such as vinegar (acetic acid), citric acid (found in limes, oranges, grapefruits, etc.), muriatic acid (hydrochloric acid, HCl), and battery acid (sulfuric acid, H_2SO_4). Common household bases include baking soda (sodium bicarbonate, $NaHCO_3$), washing soda

(Na_2CO_3), drain cleaners (sodium hydroxide, NaOH), and ammonia (NH_3).

Pure water also contains very small amounts (1×10^{-7} moles/L at 25°C) of hydronium and hydroxide ions arising from the dissociation of water molecules. However, because the concentrations of the two ions are exactly equal in pure water, there is no excess of either and the solution is classified as **neutral**.

Pure water is neither acidic nor basic because it contains equal concentrations of hydroxide and hydronium ions.

$$2\ H_2O \xrightleftharpoons{K_{eq}} H_3O^+ + HO^-$$

The pH Scale

A useful way to quantify the acid strength of an aqueous solution is the **pH scale**, where pH is defined as the negative of the logarithm of the hydronium ion concentration in moles per liter.

$$pH = -\log[H_3O^+]$$

Because the logarithm of the units of concentration cannot be calculated, pH is expressed as a unitless number. If the pH of a solution is known, the hydronium ion concentration can be calculated by taking the inverse log of the *negative* of the pH.

$$[H_3O^+] = 10^{-pH}$$

The pH scale for aqueous solutions typically ranges from 0 to 14 but can be as low as -3 or as high as 15. In a solution of pure water at 25°C, the concentrations of both OH^- and H_3O^+ are equal to 1.0×10^{-7} moles/L. When $[H_3O^+] = [OH^-]$, the solution is said to be *neutral*. If $[H_3O^+]$ exceeds that of $[OH^-]$, the solution is *acidic*, and if $[OH^-]$ exceeds that of $[H_3O^+]$, the solution is *basic* or *alkaline*. Thus the pH of an acidic solution will be less than 7 and that of a basic solution will be greater than 7. For example, if the concentration of H_3O^+ is 1×10^{-6} M, the pH is $-\log(1 \times 10^{-6}) = 6.0$. If the concentration of H_3O^+ is 1×10^{-8} M, the pH is

An aqueous solution at 25°C with pH greater than 7 is **basic** *and one with pH less than 7 is* **acidic**.

$$-\log(1 \times 10^{-8}) = 8.0$$

	$[H_3O]^+$, moles/L at 25°C	pH (25°C)
Acidic solution	$<1.0 \times 10^{-7}$	<7.00
Neutral solution	1.0×10^{-7}	7.00
Basic solution	$>1.0 \times 10^{-7}$	>7.00

Table 4-1 shows the pH of some common solutions. Notice that the pH of rainwater is significantly lower than that of pure water.

Why is this so? As rain falls through the air, carbon dioxide in the air dissolves in the water and reacts with it to form carbonic acid.

$$CO_2\,(g)\ +\ H_2O\,(l)\ \underset{\longleftarrow}{\overset{K_{eq}}{\longrightarrow}}\ H_2CO_3\,(aq)$$

Carbonic acid reacts with water to produce hydronium ions, which lowers the pH of rainwater to approximately 5.7.

$$H_2CO_3\,(aq)\ +\ H_2O\,(l)\ \underset{\longleftarrow}{\overset{K_{eq}}{\longrightarrow}}\ H_3O^+\ +\ HCO_3^-\,(aq)$$

Table 4-1: pH of Common Substances

Substance	pH
Battery acid	0.3
Gastric juice	1.0–2.0
Lemon juice	2.4
Commercial vinegar	3.0
Orange juice	3.5
Urine	4.8–7.5
Rainwater or distilled water	5.5–6.0
Milk	6.5
Saliva	6.4–6.9
Pure water	7.0
Blood	7.35–7.45
Baking soda solution	8.3
Household bleach	10.5
Household ammonia	11.5
Household lye	13.6

It is important to acknowledge the fact that for each change of one pH unit, the actual change in concentration of hydronium ions that has occurred is a very large *tenfold* change! For example, a solution of pH 5 has *10 times* the concentration of hydronium ions as a solution of pH 6.

Effects of pH in Natural Waters

The pH of natural waters has major consequences for the organisms that live there. Acidification is a particular problem, because more human inputs into natural systems are acidic than basic. A decrease in pH much below the neutral value of 7.0 can result in a variety of effects on lakes, rivers, and streams and their inhabitants.

1   *The solubility of many minerals is increased at lower pH.* In areas where there are naturally high concentrations of metal ores, the result is a release of toxic metal ions (e.g., Al^{+3}, Pb^{+2}, etc.) into the environment. Soils are typically composed of aluminosilicates (a variety of compounds containing aluminum, silicon, and oxygen), which dissolve under acid conditions to release Al^{+3} ions. Aluminum has toxic effects on many plant and animal species, including humans.

2   *A low pH dissolves the calcium from the shells of crustaceans and mollusks,* weakening them and making the animals more susceptible to physical damage as well as to predators and disease.

3   *Acid disturbs the balance in ion uptake by fish.* Fish require a balance of sodium, potassium, calcium, and chloride ions in their blood. In acidic waters, sodium ions are lost through the gills and cannot be replaced quickly enough to maintain the desired level in the blood. When the balance of ions is disturbed beyond a certain point, the fish die. Although some species of fish are more tolerant to waters with low pH than others, none can survive in waters with a pH much below 4.5.

Too much acidity in natural waters results in significant damage to fish, shellfish, and plant life.

Humans and animals can tolerate fairly large extremes in the pH of their drinking water. Consider the fact that most soft drinks have a pH between 2 and 4! Raw (untreated) public water supplies typically have a pH between 4 and 9, with the majority having a pH between 5.5 and 8.6. After treatment, most public water supplies have a pH between 6.9 and 7.4. The acceptable pH range for drinking water as dictated by the Public Health Service Act is 6.5–8.5.

The major health problem related to the pH of drinking water is that a high concentration of acid increases the solubility of metals such as lead, copper, zinc, and iron. If pipes in the plumbing system are made of these metals, water with a low pH will corrode the pipes, and the metal ions will contaminate the drinking water. Of particular concern is lead, since lead is a cumulative toxin. The ancient Romans used lead to make pipes to carry water, hence the word "plumbing" is derived from the Latin word for lead, *plumbum* (chemical symbol, Pb). The pipes in many water distribution systems built before 1986 contain lead, and if the pH is less than 7, lead leaches out of the pipes and into the drinking water. Thus it is in the best interest of the public health if the pH of drinking water is not acidic.

If the pH of a drinking water supply is too acidic, lead water pipes will dissolve into the water supply and pose a lead poisoning hazard.

TECHNIQUES FOR MEASURING pH

Electrochemical Measurement of pH

The pH electrode compares the concentration of H_3O^+ outside the electrode to a standard solution of known H_3O^+ concentration inside the electrode.

Most laboratory pH measurements are carried out using a pH meter. The pH meter works by using an electrical circuit that contains a glass electrode, an external reference electrode, and a voltmeter that gives a measure of the electrical potential in the circuit. The glass electrode consists of a current-carrying wire that dips into a solution of known pH (see Figure 4-1). This solution is encapsulated in a thin glass membrane made of aluminosilicate glass (see Figure 4-2). The molecular structure of the glass is such that oxygen atoms comprise much of the surface of the glass, with the backbone of the molecule consisting of aluminum and silicon atoms.

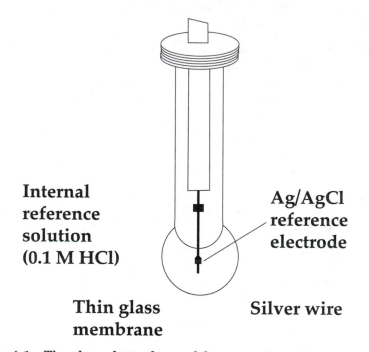

Internal reference solution (0.1 M HCl)

Ag/AgCl reference electrode

Thin glass membrane

Silver wire

Figure 4-1: The glass electrode, used for measuring pH.

Because oxygen atoms are electronegative and have unshared pairs of electrons, they will have a partial negative charge and will thus be "sticky" to positively charged species like protons. When the glass electrode is dipped into an acidic sample to measure the pH, the protons in the sample stick to the outside of the glass membrane. In order to maintain neutrality of the membrane, some of the protons that are stuck to the surface on the inside of the glass membrane desorb from the surface. The consequence is a change in the pH of the inner solution, which results in an electrical potential. When the electrode is dipped into a basic sample, the protons adhering to the external surface of the glass electrode are removed by the hydroxide ions in solution. Again, to neutralize the charge on the membrane, protons from the inner solution adhere to the inner surface, thereby causing a change in electrical potential within the electrode which is then translated into a pH reading by the meter. Before the reading can be trusted, the meter must first be standardized by placing the electrodes into buffer solutions of known pH and adjusting the meter to read the known value.

Glass electrode placed in an acid solution (low pH)

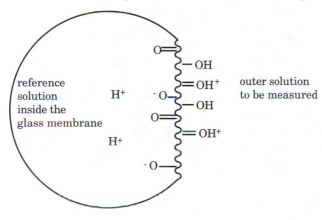

Glass electrode placed in a basic solution (high pH)

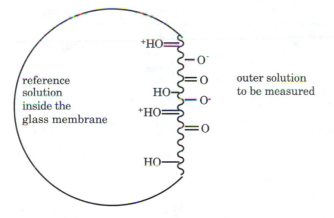

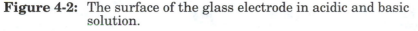

Figure 4-2: The surface of the glass electrode in acidic and basic solution.

An external reference electrode, usually Ag/AgCl, is coupled to the glass electrode to complete the circuit, and the voltmeter gives a measure of the electrochemical potential of the cell, as defined by the Nernst equation. For this system, the cell voltage, E_{cell}, is given by:

$$E_{cell} = \text{constant} - 0.059 \ pH$$

where the constant is characteristic of the particular electrochemical cell. The pH meter is calibrated by placing the glass electrode in solutions of known pH and using a voltmeter to measure E_{cell}. This is the equivalent of making a straight line plot (standard curve) of E_{cell} vs. pH and allows calculation of the pH of an unknown sample by comparison to the standard curve.

Colorimetric Determination of pH

The chemical method used for determining pH involves addition of a drop of a solution of an acid-base indicator to the sample of interest. Alternatively, pH indicator papers can be used, on which a drop of the sample is placed and the color noted. Indicators are large organic molecules that change color depending on the pH of the solution. There is a range of pH over which the color changes for each indicator (usually around 1.5 pH units), with intermediate colors observed for pH values in the middle of this range. Consider the indicator cresol red as an example. At pH values below 7.0, a solution of the indicator is yellow, and above pH 8.8, the solution is red. At intermediate pH values, the solution will be varying shades of orange, depending on the exact pH. Many different indicators are available, thus covering the entire pH range (see Table 4-2).

Acid-base indicator solutions can also be used to determine the pH of a sample.

Table 4-2: Indicators Used to Determine pH

Indicator	Color Change (acidic ——> basic)	pH Range in Which Color Change Occurs
Malachite green	Yellow ——> green	0.2–1.8
Thymol blue	Red ——> yellow Yellow ——> blue	1.2–2.8 8.0–9.6
Methyl orange	Red ——> yellow	3.2–4.4
Bromcresol green	Yellow ——> blue	3.8–5.4
Methyl red	Red ——> yellow	4.8–6.0
Bromothymol blue	Yellow ——> blue	6.0–7.6
Cresol red	Yellow ——> red	7.0–8.8
Phenolphthalein	Colorless ——> pink	8.2–10.0
Thymolphthalein	Colorless ——> blue	9.4–10.6
Alizarin yellow	Yellow ——> red	10.1–12.0

LABORATORY PROCEDURES: MEASURING pH

Objective: To determine the pH of a water sample. Background information on pH can be found on pages 52–55.

Method 1: Measuring Sample pH Using a pH Meter

The pH meter is the instrument of choice for precise measurement of both pH and total alkalinity. The glass membrane electrode used in the pH meter is selective for hydronium ions (H_3O^+) by detecting a difference in the concentration of H_3O^+ between the solution inside the electrode and the solution outside the electrode. The procedure for measuring pH is given below.

Calibration of the pH Meter

In order to get correct readings, the pH meter must first be calibrated using solutions of known pH. Buffer solutions of known pH work well for this purpose. Calibration buffers are chosen with pH values that bracket the expected pH of the sample solution, usually pH 4.0 and 7.0 or 7.0 and 10.0. While the exact instructions for use of the pH meter vary slightly depending on the model, the general standardization procedure is outlined below.

Note

All glassware used in this experiment should be acid-washed and well rinsed according to the procedures on page 25.

The standard procedure for transferring the electrode from one solution to another is as follows: Remove the electrode from the solution and rinse it *well* with deionized water, using your wash bottle. Shake the excess water off the electrode and place it in the new solution.

Note

Do not touch the bottom of the electrode, as this will depolarize it and give inaccurate readings.

Note

Do not let the electrode dry out. Be sure it is immersed in a solution except when transferring between beakers.

1 Turn the meter on. Rinse the electrode with deionized water, shake it dry, and place it in a beaker containing the first buffer solution at pH 7.00. If you are using a magnetic stirrer, check to be sure there is a magnetic stir bar in the beaker and adjust the stirrer so the stir bar spins slowly. Wait for the reading to stabilize, then adjust the reading to pH 7.00.

2 Rinse the electrode with deionized water, shake it dry, and place it in a beaker of the second standard buffer solution (pH 4.00 or 10.01, depending on the expected pH of the

samples). Allow the reading to stabilize and then adjust it to the appropriate pH for the buffer solution used.

3   Check the calibration by measuring the pH of the 7.00 standard. To do so, place the electrode in the solution, wait for a stable reading, and make a note of the number. If the reading is more than 0.05 pH units different than 7.00, repeat the standardization procedure, steps 1 through 3.

4   At this point the meter is ready to use.

Measuring Sample pH

It is standard procedure to measure pH on an *unfiltered* sample.

Note

All glassware used in this experiment should be acid-washed and well rinsed according to the procedures on page 25.

Note

1   Take the pH of your sample by filling a *clean* 100-mL beaker with enough sample to immerse the (clean) electrode into the beaker, stir the solution, allow the reading to stabilize for about 2 minutes, and read the display to obtain the pH of your sample. Note this value in your notebook.

2   If the pH of your sample is below 4.5, there is no need to determine total alkalinity, since the sample will have *no* acid neutralizing capacity. If the pH is *above* 4.5, proceed to the total alkalinity determination.

If you dispose of your sample at this point, you have not added any chemicals, so it may be disposed of down the drain.

Waste
Disposal

Method 2: Measuring Sample pH Using pH Paper

A fast, inexpensive, and convenient technique for determination of the pH of a sample is to use pH paper. The paper has been treated with a solution of an acid-base indicator that has a particular color at a given pH. Each paper comes with a calibrated scale for color comparison.

1 Take the pH of your sample by placing a drop of the sample onto a wide-range (1–14) pH paper. Use the color comparison chart to determine the pH to the nearest pH unit and record this number.

2 Choose the appropriate narrow range pH paper and place a drop of the sample on the paper. Use the color comparison chart to determine the pH to the nearest tenth of a pH unit and record this number in your lab notebook.

3 If the pH of your sample is below 4.5, there is no need to determine total alkalinity, since the sample will have *no* acid neutralizing capacity. If the pH is *above* 4.5, proceed to the total alkalinity determination.

THE ROLE OF ALKALINITY IN WATER QUALITY

In pure water, addition of even small quantities of a strong acid or strong base results in a very dramatic pH change. For example, if just one drop (0.05 mL) of a 0.1 molar solution of HCl is added to 250 mL (about a cup) of pure distilled water, the pH will change from neutral (7.0) to a quite acidic 4.7!

Such significant pH changes from very small inputs of acid can have a devastating effect on an ecosystem. Fortunately, most natural waters are protected from such drastic change by the presence of ionic compounds in the water that react quickly with acid to **buffer** the system and thereby resist a change in pH. This capacity of a solution to neutralize acid is called **acid neutralizing capacity (ANC)** or **total alkalinity**.

Neutralization Reactions

The reactions of basic compounds with acids to form harmless species are called **neutralization reactions** and arise by combination of free hydronium ions in the water with basic species such as hydroxide, carbonate, and bicarbonate to form harmless compounds. These types of reactions are usually rapid and proceed essentially to completion.

The compounds that are responsible for total alkalinity in natural waters are typically mineral salts containing carbonate, bicarbonate, and hydroxide anions, with calcium and magnesium as the positively charged counterions (see Table 4-3). The neutralization reactions that are responsible for the buffering action involve the anions of these compounds and include the following:

Acid inputs into an ecosystem can be neutralized by the presence of minerals containing bases such as bicarbonate and carbonate.

Carbonate:

$$H_3O^+(aq) \; + \; CO_3^{-2}(aq) \longrightarrow HCO_3^-(aq) \; + \; H_2O\,(l)$$

Bicarbonate:

$$H_3O^+(aq) \; + \; HCO_3^-(aq) \longrightarrow H_2CO_3(aq) \; + \; H_2O\,(l)$$

Hydroxide:

$$H_3O^+(aq) \; + \; HO^-(aq) \longrightarrow 2\,H_2O\,(l)$$

Table 4-3: Common Minerals with Buffering Capacity

Mineral	Chemical Composition
Calcite	$CaCO_3$
Magnesite	$MgCO_3$ (rare)
Dolomite	$CaCO_3 \cdot MgCO_3$
Brucite	$Mg(OH)_2$ (rare)

Knowledge of the pH of a body of water is incomplete without a knowledge of total alkalinity. Two samples of identical pH, but with different total alkalinities will behave very differently on addition of acid to the sample. The solution with high total alkalinity will be able to resist a change in pH on addition of acid up to the limit of its buffering capacity, while the solution with low total alkalinity will show a change in pH soon after addition of even small amounts of acid.

A water supply with high total alkalinity is resistant to changes in pH.

The neutralization of acidic industrial discharges with bases such as calcium hydroxide ($Ca(OH)_2$, also known as lime) and sodium carbonate (Na_2CO_3, also known as soda ash) is a common practice. Neither the calcium or sodium ions contribute to the neutralization process—it is the negative counterion, hydroxide or carbonate, that is responsible for the neutralization process.

TECHNIQUES FOR MEASURING TOTAL ALKALINITY

Measuring Total Alkalinity by Titration

The total alkalinity or acid neutralizing capacity of a sample is determined by **titration** of the sample with a strong acid of a known concentration to a specified pH, i.e., determining exactly how much acid the solution was able to neutralize. A titration is simply a dropwise addition of one solution to another to a chosen endpoint. The chemical reaction occurring is:

$$CaCO_3 \text{ (aq)} + H_2SO_4 \text{ (aq)} \longrightarrow H_2CO_3 \text{ (aq)} + CaSO_4 \text{ (aq)}$$

Excluding spectator ions, it is easier to see that carbonate is acting as a base (a proton acceptor) and is being neutralized by the acid H_3O^+.

$$CO_3^{-2} \text{ (aq)} + 2 H_3O^+ \text{ (aq)} \longrightarrow H_2CO_3 \text{ (aq)} + 2 H_2O \text{ (l)}$$

The pH value commonly used as the endpoint for the total alkalinity titration is 4.5. The endpoint can be detected in one of two ways: (1) by monitoring the course of the titration with a pH meter or, (2) by using a colored indicator compound that changes color at the correct pH.

The **equivalence point** and the **endpoint** are not necessarily the same thing. The endpoint is simply the point at which there is some physical evidence that the neutralization reaction has just gone to completion. This may be a color change, a pH value specified by a method, or another observable change in the solution. The equivalence point is defined as the point in the titration at which the moles of hydronium ions added exactly equals the moles of **base equivalents** initially present in the solution. It is necessary to distinguish between the moles of base present and the moles of base equivalents because different bases can neutralize different numbers of acidic protons. For example, the hydroxide ion can only neutralize one acidic proton, but the carbonate ion can neutralize **two** acidic protons because it has two sites that can bind to H⁺ ions.

*The **endpoint** of a titration is not necessarily the same as the **equivalence point**.*

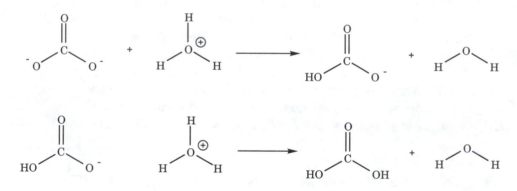

The reporting of total alkalinity is often simplified by assuming all acid neutralizing capacity is due to the presence of $CaCO_3$, and is thus reported as milligrams of $CaCO_3$ per liter. Although alkalinity is frequently caused by substances other than $CaCO_3$, the net effect in terms of acid neutralizing capacity is essentially the same, so while this reporting procedure is not exactly precise, the value of the acid neutralizing capacity is not compromised.

Expected Total Alkalinity Values

The range of total alkalinity values for natural waters is typically between 30 and 500 mg of $CaCO_3$/L, with higher values occurring in regions that have alkaline soils. Rainwater usually has very little total alkalinity (<10 mg/L) since it has contact with few minerals. Surface waters generally have total alkalinities less than 200 mg/L, whereas groundwater total alkalinities are frequently much higher, sometimes over 1000 mg/L due to increased pressure of CO_2 from microbial degradation of organic matter underground. The CO_2 reacts with water to form bicarbonate and acid.

Surface waters typically have much lower alkalinities than groundwater.

$$CO_2 \text{ (g)} + 2\,H_2O \text{ (l)} \longrightarrow HCO_3^- \text{ (aq)} + H_3O^+ \text{ (aq)}$$

The acid (H_3O^+) formed in this reaction reacts with calcium or magnesium carbonate present in the surrounding rock to dissolve these compounds and produce bicarbonate, which is the main contributor to total alkalinity.

$$H_3O^+(aq) \ + \ CaCO_3 \ (s) \longrightarrow H_2O \ (l) \ + \ HCO_3^- \ (aq) \ + \ Ca^{+2} \ (aq)$$

Seawater has a total alkalinity of typically 200-500 mg/L, due to the high concentrations of dissolved bicarbonate ions.

LABORATORY PROCEDURES: MEASURING TOTAL ALKALINITY

Objective: To determine the acid neutralizing capacity (alkalinity) of the water supply. Background information on alkalinity can be found on pages 62–65.

It is standard procedure to measure total alkalinity on an *unfiltered* sample.

Note

All glassware used in this experiment should be acid-washed and well rinsed according to the procedures on page 25.

Note

Method 1: Total Alkalinity Using a pH Meter and a Buret to Titrate

1   Calibrate the pH meter according to the instructions on page 59.

2   Prepare your buret by filling it to the top of the graduated area with standard 0.0100 M sulfuric acid.

The concentration of the acid prepared for your class will not be exactly 0.0100 M. Be sure to write down the EXACT concentration from the bottle you used to fill your buret.

Note

3   Check that there are no air bubbles in the buret tip. If there are, allow a few milliliters of acid to flow out. You may need to tap the buret to dislodge the bubble.

4   In your lab notebook, make a table with two columns to monitor the pH as you titrate the sample with acid. A sample table is shown below. When you read the buret, always have your eyes at the same height as the liquid and read the value at the bottom of the meniscus to one decimal (e.g., 10.1) and estimate the second decimal.

Sample Data Table for Buret Titration

pH	Buret Reading (mL)	Amount of Acid Added (mL)
7.43	49.98	0.00
7.12	47.25	2.73
6.86	44.78	2.47
.	.	.
.	.	.
etc.	etc.	etc.

5 Measure 100.0 mL of sample with a graduated cylinder into a clean 250-mL beaker. Immerse the (clean) electrode into the beaker, stir the solution, allow the reading to stabilize for about 2 minutes, and read the display to obtain the initial pH of your sample. Note this value in the table.

6 Swirling the sample gently, titrate the solution with 0.0100 M sulfuric acid (H_2SO_4). The chemical reaction you are monitoring is:

$$CaCO_3 \text{ (aq)} + H_2SO_4 \text{ (aq)} \longrightarrow H_2CO_3 \text{ (aq)} + CaSO_4 \text{ (aq)}$$

Excluding spectator ions:

$$CO_3^{-2} \text{ (aq)} + 2\, H_3O^+ \text{ (aq)} \longrightarrow H_2CO_3 \text{ (aq)} + 2\, H_2O \text{ (l)}$$

7 In your notebook, record the pH, the buret reading, and the amount of acid added at regular intervals. As the pH approaches 5.0, proceed more slowly, recording the pH at smaller volume intervals. Consultation with your labmates will help you predict the approximate amount of acid necessary to titrate the sample to the pH 4.5 endpoint. When these procedures are complete, rinse the electrode and place it back into a beaker of deionized water.

Waste Disposal

The solutions that were titrated are approximately the same pH as orange juice and may be disposed of down the drain.

Calculations

From the volume of acid used to titrate the sample, calculate the Total Alkalinity in mg/L as $CaCO_3$. This is a four-step process:

1 Convert milliliters of H_2SO_4 used to titrate to pH 4.5 to moles of H_2SO_4.

2 Use the mole ratio of reactants in the reaction of $CaCO_3$ and H_2SO_4 to determine the moles of $CaCO_3$ that must have been present.

3 Convert moles of $CaCO_3$ to mg of $CaCO_3$.

4 Use the volume of the sample titrated to obtain mg of $CaCO_3$ per liter of sample.

Plotting a Titration Curve

A plot of the volume of acid added versus pH provides information about the buffering ability of the sample, as well as the K_b of the base HCO_3^-. In this assignment you will plot a titration curve from the alkalinity titration data obtained.

1 On a piece of graph paper (or use a computer graphing program to generate a plot), graph the pH of the solution (y-axis) vs. the volume of acid added (x-axis). Label the buffering region on the graph.

2 Write the equation for the neutralization reaction of calcium carbonate with sulfuric acid.

3 The **endpoint** chose for this titration was pH 4.5. The pH at the exact **equivalence point** (see p. 64) will be slightly different from pH 4.5. Calculate the pH at the equivalence point for your sample, assuming that all alkalinity stems from the presence of $CaCO_3$. Do this calculation for three hypothetical samples as well, having alkalinities of 30, 100, and 500 mg/L, respectively. The K_a for the dissociation of carbonic acid is 4.3×10^{-7}.

4 Label the equivalence point for your sample on the graph.

5 Why do you think pH 4.5 was chosen as the endpoint for an alkalinity titration?

6 Using a dotted line, draw in the part of the titration curve (after the equivalence point) for which data are not available.

Method 2: Total Alkalinity Using a pH Meter and a Hach Titrator

> ⚠️ **Note** All glassware used in this experiment should be acid-washed and well rinsed according to the procedures on page 25.

1 Calibrate the pH meter according to the instructions on page 59.

2 In your lab notebook, make a table with two columns to monitor the pH as the sample is titrated with acid. A sample table is shown as follows.

Sample Data Table for Titration with Hach Titrator

pH	Amount of Acid Added (digits)
7.43	0
7.12	12
6.86	25
.	.
.	.
etc.	etc.

3 Measure 100.0 mL of sample with a graduated cylinder into a clean 250-mL beaker. Immerse the (clean) electrode into the beaker, stir the solution, allow the reading to stabilize for about 2 minutes, and read the display to obtain the initial pH of your sample. Note this value in the table.

4 Prepare the Hach titrator by turning the delivery knob to remove any air in the tip of the small plastic tube. Wipe off any drops of acid with a Kimwipe® and zero the digit counter.

5 With the sample stirring slowly with a magnetic stir bar, use the Hach titrator to add 0.8 M (1.6 N) sulfuric acid (H_2SO_4) to the solution by turning the delivery knob on the titrator very slowly while stirring the solution gently with the plastic tip of the titrator.

> ⚠️ **Note** If the sample has very low total alkalinity such as is typically found for rainwater or snow, use a more dilute acid. The Hach 0.08 M (0.16 N) H_2SO_4 cartridges work well for this purpose.

The chemical reaction you are monitoring is:

$$CaCO_3 \text{ (aq)} + H_2SO_4 \text{ (aq)} \longrightarrow H_2CO_3 \text{ (aq)} + CaSO_4 \text{ (aq)}$$

Excluding spectator ions:

$$CO_3^{-2} \text{ (aq)} + 2\,H_3O^+ \text{ (aq)} \longrightarrow H_2CO_3 \text{ (aq)} + 2\,H_2O \text{ (l)}$$

6 In your notebook, record the pH and the number of digits of acid added at regular intervals (about every 10 digits on the titrator). As the pH approaches 5.0, add acid more slowly, recording the pH at smaller volume intervals. Consultation with your labmates will also help you predict the approximate amount of acid necessary to titrate the sample to the pH 4.5 endpoint. When these procedures are complete, rinse the electrode and place it back into a beaker of deionized water.

7 Calculate the total alkalinity (in mg/L as $CaCO_3$), using the digits on the titrator. For a 100.0-mL sample using 0.8 M H_2SO_4 as the titrant, the total alkalinity is given by:

$$\text{Total alkalinity} = \text{number of digits to pH 4.5}$$

Questions to Answer in the Lab Notebook

1 The foregoing formula provides the concentration of total alkalinity in mg of $CaCO_3$ per liter. Calculate the corresponding concentration of $CaCO3$ in moles per liter.

2 How many moles of H_2SO_4 did you add in the titration?

3 How many milliliters of H_2SO_4?

4 How many "digits" are there in a milliliter?

Plotting a Titration Curve

A plot of the volume of acid added versus pH provides information about the buffering ability of the sample, as well as the K_b of the base HCO_3^-. In this assignment you will plot a titration curve from the alkalinity titration data obtained.

1 On a piece of graph paper (or use a computer graphing program to generate a plot), graph the pH of the solution (*y*-axis) vs. the volume of acid added (*x*-axis). Label the buffering region on the graph.

2 Write the equation for the neutralization reaction of calcium carbonate with sulfuric acid.

3 The **endpoint** chose for this titration was pH 4.5. The pH at the exact **equivalence point** (see p. 64) will be slightly different from pH 4.5. Calculate the pH at the equivalence point for your sample, assuming that all alkalinity stems from the presence of $CaCO_3$. Do this calculation for three hypothetical samples as well, having alkalinities of 30, 100, and 500 mg/L, respectively. The K_a for the dissociation of carbonic acid is 4.3×10^{-7}.

4 Label the equivalence point for your sample on the graph.

5 Why do you think pH 4.5 was chosen as the endpoint for an alkalinity titration?

6 Using a dotted line, draw in the part of the titration curve (after the equivalence point) for which data are not available.

Method 3: Total Alkalinity Using an Indicator Titration

A fast, inexpensive, and convenient technique for determination of the total alkalinity of a sample is to titrate the sample with acid in the presence of an acid-base indicator, a substance that has different colors at different pH values. Bromcresol green is the indicator of choice for the total alkalinity titration, since it changes from blue above pH 5.4 to yellow below pH 3.8. At pH 4.5, the indicator is green.

All glassware used in this experiment should be acid-washed and well rinsed according to the procedures on page 25.

Note

1   Prepare your buret by filling it to the top of the graduated area with standard 0.0100 M sulfuric acid.

The concentration of the acid prepared for your class will not be exactly 0.0100 M. Be sure to write down the **EXACT** concentration from the bottle you used to fill your buret.

Note

2   Check that there are no air bubbles in the buret tip. If there are, allow a few milliliters of acid to flow out. You may need to tap the buret to dislodge the bubble.

3   Take the initial volume reading from the buret. When you read the buret, always have your eyes at the same height as the liquid and read the value at the bottom of the meniscus to one decimal (e.g., 10.1) and estimate the second decimal.

4   Prepare the sample for the total alkalinity determination by filling a clean 250-mL Erlenmeyer flask with 100 mL of the sample (measured with a graduated cylinder). Add a drop or two of bromcresol green indicator. If the pH of your sample is below 4.5 (indicator is yellow), there is no need to determine alkalinity, since the sample will have *no* acid neutralizing capacity. If the pH is *above* 4.5, proceed to the next step.

5   Swirling the sample gently, titrate the solution with 0.0100 M sulfuric acid (H_2SO_4). The chemical reaction you are monitoring is:

$$CaCO_3 \text{ (aq)} \ + \ H_2SO_4 \text{ (aq)} \longrightarrow H_2CO_3 \text{ (aq)} \ + \ CaSO_4 \text{ (aq)}$$

Excluding spectator ions:

$$CO_3^{-2} \text{ (aq)} \ + \ 2 \, H_3O^+ \text{ (aq)} \longrightarrow H_2CO_3 \text{ (aq)} \ + \ 2 \, H_2O \text{ (l)}$$

6   Consultation with your labmates will help you predict the approximate amount of acid necessary to titrate the sample to the pH 4.5 endpoint, where the indicator changes color from blue to green. If you pass the endpoint, the solution will be yellow and you should redo the titration.

Calculations

1   From the volume of acid used to titrate the sample, calculate the total alkalinity in milligrams per liter as $CaCO_3$. This is a four-step process:

    a   Convert mL of H_2SO_4 to moles of H_2SO_4

    b   Use the mole ratio of reactants in the reaction of $CaCO_3$ and H_2SO_4 to determine the moles of $CaCO_3$ that must have been present initially.

    c   Convert moles of $CaCO_3$ to mg of $CaCO_3$.

    d   Use the volume of the sample titrated to obtain mg of $CaCO_3$ per liter of sample.

2   The **endpoint** chose for this titration was pH 4.5. The pH at the exact **equivalence point** (see p. 64) will be slightly different from pH 4.5. Calculate the pH at the equivalence point for your sample, assuming that all alkalinity stems from the presence of $CaCO_3$. Do this calculation for three hypothetical samples as well, having alkalinities of 30, 100, and 500 mg/L, respectively. The K_a for the dissociation of carbonic acid is 4.3×10^{-7}.

Chapter 5

Anions in Natural Waters

The most common inorganic anions found in natural waters between pH 7-9 include bicarbonate (HCO_3^-, the anion responsible for alkalinity, see page 62), hydrogen sulfate (HSO_4^-, often referred to as just SO_4^{-2} or sulfate), and chloride (Cl^-). The ionic species nitrate (NO_3^-), dihydrogen phosphate ($H_2PO_4^-$, often referred to as just PO_4^{-3} or phosphate), and fluoride (F^-) occur in somewhat lower concentrations. Although there are other anionic components in natural waters, these common anions are in high enough concentration to be easily monitored and can readily be used as indicators of sources of pollution.

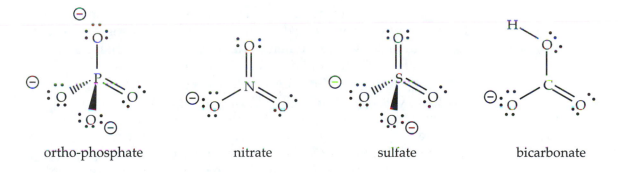

| ortho-phosphate | nitrate | sulfate | bicarbonate |

NITRATE AND PHOSPHATE: THE NUTRIENT ANIONS

Nitrate and phosphate are considered to be **nutrients**, essential for plant growth. Fertilizer usually consists of a mixture of the ionic compounds ammonium nitrate (NH_4NO_3), potassium nitrate (KNO_3), and ammonium dihydrogen phosphate ($NH_4H_2PO_4$). These ionic substances dissolve in rainwater or irrigation water and are taken up and utilized by plants. If too much fertilizer is applied or if it rains hard after an application of fertilizer, the runoff from fields can be a major source of nutrient pollution of both surface water and groundwater. Nitrate is the most common agricultural pollutant because it is very water soluble and travels freely with runoff water. It is also a health hazard in drinking water at levels greater than 10 mg/L. Prevention of nitrate pollution is aided by using only as much fertilizer as is needed and by fertilizing only during the growing season so the plants take up the nutrients immediately.

Agricultural activities contribute nutrients to an ecosystem, often overloading the ability of the system to process the nutrients.

Excess phosphate can also pose serious environmental problems. While nitrogen, carbon, and phosphorus are the major classes of nutrients required for the growth of aquatic plants, it is usually the supply of phosphorus that limits the growth of algae in water. When excess phosphorus in the form of phosphate (PO_4^{-3}) and nitrogen in the form of nitrate (NO_3^-) or ammonia (NH_4^+) enter a water supply, either from agricultural runoff or from municipal waste discharges containing phosphates, **eutrophication**, or nutrient enrichment, of lakes and streams occurs. Although eutrophication is a natural phenomenon associated with aging of a lake, the process can be greatly accelerated by contributions of pollutants from human sources. A eutrophic body of water can produce massive **algal blooms**. The algae present in the contaminated body of water grow rapidly in the presence of excess nutrients. While they are growing, they produce oxygen by photosynthesis during daylight hours. When it is dark, they change to a different metabolic process that utilizes available oxygen in the water.

When the algae die, oxygen production ceases and bacterial decomposition occurs. The decomposition process uses up much of the available dissolved oxygen in the water, and the fish and other aquatic life suffocate and die. **Red tides**, recognized by a reddish tint in the water of affected areas, are algal blooms of the marine organism *Gonyaulax sp.* When this dinoflagellate grows, it produces toxins similar to the botulism toxin. These toxins poison fish and may even persist in shellfish. Humans eating shellfish from a red tide area have been poisoned by this toxin.

Red tides are algal blooms caused by excessive nutrients such as nitrate and phosphate in the water.

In certain parts of the country, the major source of nutrient pollution is runoff from feedlots. Animal wastes contain significant amounts of ammonium (NH_4^+) and nitrate (NO_3^-) ions, as well as high concentrations of organic matter. The average steer excretes 27 kg of waste daily, eighty-five percent of which is water. The remaining 15% contains 180 g of nitrogen as nitrate or ammonia, as well as fecal coliform bacteria. When many cattle are confined to a small area, the potential for pollution from the runoff is quite high. Not only are nearby streams contaminated, but the land the feedlots are located on becomes very concentrated in nitrogen. Many states now require that feedlots install drainage to divert feedlot runoff into settling ponds or lagoons for treatment before allowing the water to flow back into a river or stream. Of course, the positive side of this story is that the feedlot waste is very good fertilizer and is frequently spread on nearby fields to enrich the soil.

A minor source of nitrate in surface waters (but the major source in rainwater or snow) is combustion reactions in which nitrogen in the air combines with oxygen at high temperatures to form nitrogen oxides. Power plants and automobiles are major sources of nitrogen oxides.

Nitrogen oxides from combustion of fossil fuels are only a minor source of nitrate in surface waters

$$N_2 (g) + O_2 (g) \xrightleftharpoons{\text{high temperature}} 2\, NO (g)$$

$$NO (g) + 1/2\, O_2 (g) \rightleftharpoons NO_2 (g)$$

The compounds NO, NO_2, and N_2O_4, commonly abbreviated NO_x, react with water and oxygen in the atmosphere to form nitric acid, HNO_3. During a rain or snow storm, nitric acid is washed out of the atmosphere and ends up in the surface waters.

SULFATE: GEOLOGICALLY AVAILABLE

Many minerals occur as sulfates or as sulfides that are transformed into sulfates on exposure to air. Natural weathering of these minerals contributes to sulfate concentrations in surface waters. High concentrations of sulfate in natural waters is most commonly associated with mining activities or the presence of the common mineral gypsum ($CaSO_4$). Table 5-1 shows some common sulfate and sulfide minerals and their chemical formulas.

Table 5-1: Common Sulfate and Sulfide Minerals

Mineral	Chemical Formula
Gypsum	$CaSO_4 \cdot 2\, H_2O$
Pyrite	FeS_2
Galena	PbS
Sphalerite	ZnS

The oxidation of sulfide minerals to sulfate is accompanied by the formation of sulfuric acid. For pyrite, a common iron sulfide ore, the chemical reaction that takes place is:

$$4\, FeS_2 (s) + 14\, O_2 (g) + 4\, H_2O (l) \rightleftharpoons 4\, Fe^{+2} (aq) + 8\, H^+ (aq) + 8\, SO_4^{-2} (aq)$$

In the presence of Fe^{+3}, pyrite can also react with water to form sulfuric acid.

$$FeS_2 (s) + 14\, Fe^{+3} (aq) + 8\, H_2O (l) \rightleftharpoons 15\, Fe^{+2} (aq) + 2\, SO_4^{-2} (aq) + 16\, H^+ (aq)$$

A minor source of sulfate in surface waters (but the major source in rainwater or snow) is the combustion of fuels. Both coal and oil contain elemental sulfur (S) or reduced sulfur compounds such as H_2S that are transformed into the gaseous sulfur oxides SO_2 and SO_3 when the fuel is burned. These sulfur oxides react with water in the atmosphere to form sulfuric acid that "rains out" in the next precipitation event.

Mining operations are responsible for a large influx of sulfate into surface waters

Sulfate is not particularly toxic but may make water taste bitter at concentrations over 500 mg/L and can cause diarrhea at concentrations over 1000 mg/L. A more significant problem with high levels of sulfate is the potential for the formation of calcium sulfate scale on high pressure boilers. To prevent problems from excessive scale formation, concentrations of sulfate in a water supply should be less than 250 mg/L.

COLORIMETRIC METHODS OF ANALYSIS

There are a variety of methods used for measuring the concentration of the constituents of a water sample. One of the more useful methods involves reacting the chemical species of interest with reagents that produce a colored compound, a technique called **colorimetric** analysis. Colorimetric analyses can be used to measure the concentration of nitrate, phosphate, and sulfate ions in solution. This section begins with a general introduction to the technique, followed by detailed discussion of the nitrate, phosphate, and sulfate analysis.

Theory of Light Absorption

The characteristic energy absorbed by molecules can be used to quantify the concentration of a molecule in solution.

In order to more fully understand this phenomenon at the molecular level, it is necessary to consider how molecules absorb light energy. A simplified way to understand it is to think of a molecule having two possible energy states, a **ground state** and **excited state**. The ground state is at a lower energy than the excited state, but with an input of energy it is possible to transform the ground state molecule into the excited state. For some molecules, normal room light in the **visible** region of the electromagnetic radiation spectrum has enough energy to cause this transformation (see Figure 5-1). It is these molecules that look colored to our eye because they transmit light in the visible range.

———— Molecule* (excited state)

Energy absorbed

E = hν

———— Molecule (ground state)

Figure 5-1: Colorimetric analyses rely on the absorption of energy in the visible range of the electromagnetic radiaion spectrum that transform molecules from the ground state to the excited state.

The separation between molecular energy states is characteristic of a particular molecule. This energy gap (E) is directly proportional to the **frequency** (v) of light absorbed by the molecule and inversely proportional to the **wavelength** (λ) of light, as related by the speed of light (c) and Planck's constant, h.

$$E = h\nu = \frac{hc}{\lambda}$$

The Spectrophotometer

Colorimetric analyses are carried out using a **spectrophotometer**, an instrument that passes a narrow range of wavelengths of light through the sample and measures the amount of light that is transmitted. A schematic of a spectrophotometer is shown in Figure 5-2.

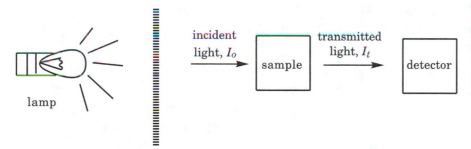

The Spectronic 21 is a common and easy-to-use spectrophotometer.

Figure 5-2: The spectrophotometer consists of a visible light source with a monochromator that allows the analyst to select a small range of wavelengths for the analysis. Light passing through the sample is absorbed by the species of interest and the intensity of the transmitted light is detected.

The energy that is absorbed from the incident light transforms the molecule from the ground state to an excited state. The result is that not all of the light that was directed at the sample will be transmitted through to the other side (see Figure 5-3).

Figure 5-3: Only a fraction of the incident light is actually transmitted through the sample.

For a simple analogy, think about how a flashlight beam might behave if you were to shine it through a glass of apple juice. On the other side of the glass, the light would be dimmer than the flashlight beam by itself. The **transmittance, T,** of the solution is just the ratio of transmitted light to incident light:

$$T = \frac{I_t}{I_o}$$

where I_o is the intensity of the incident light and I_t is the intensity of the light that was transmitted through the sample.

The wavelength chosen for a particular colorimetric analysis is the one at which the maximum amount of light is absorbed for a given concentration. A typical plot of wavelength versus light absorbed is shown in Figure 5-4.

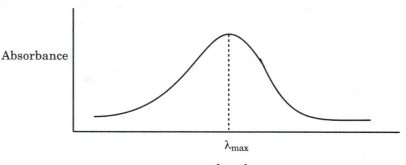

Figure 5-4: The amount of light absorbed by a molecule in solution varies with wavelength. Analyses of colored species are typically carried out at λ_{max}, the wavelength of maximum absorbance.

Relating the Amount of Light Absorbed to Concentration: Beer's Law

Before a colorimetric method can be used to determine the concentration of a species in solution, the spectrophotometer must be **standardized** to determine the relationship between the amount of light absorbed by a given quantity of the substance being measured. It is possible to measure either **absorbance (A)** of a sample or transmittance, **T**, where the relationship between absorbance and transmittance is a logarithmic one.

$$A = -\log T$$

Absorbance is linearly related to concentration by **Beer's Law**:

$$A = \varepsilon bc$$

where A is the absorbance (a unitless number), b is the path length of the light through the sample (typically ~1 cm for a spectrophotometer), c is the concentration of the substance in solution, and ε (the Greek letter *epsilon*) is a constant that is specific for a particular substance and is determined experimentally for each species analyzed. The higher the concentration of the substance in solution, the greater the absorbance that is measured by the instrument. This relationship is typically linear only over a limited range of concentrations, deviating from linearity at higher concentrations because of interactions between absorbing species. A typical Beer's Law plot, also known as a **standard curve**, is shown in Figure 5-5.

The amount of light absorbed by a substance is proportional to its concentration.

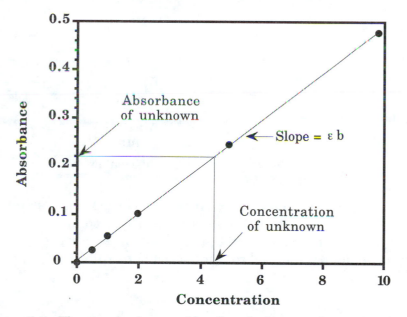

Figure 5-5: The standard curve describes the linear relationship between absorbance and concentration.

To construct a standard curve, it is necessary to make up solutions of known concentrations, measure their absorbances using the spectrophotometer, and plot the results as in Figure 5-5. The slope of the resulting line is εb. The concentration of the species of interest in an unknown solution can then be determined by using this value of the slope and Beer's Law to calculate a concentration from an absorbance reading.

Colorimetric Method for the Determination of Nitrate in Water

Nitrogen exists in a variety of ionic forms in natural waters, including nitrate (NO_3^-), nitrite (NO_2^-), and ammonium (NH_4^+). Nitrate is the most highly oxidized form of nitrogen and is the most stable form of nitrogen in an oxidizing environment like our atmosphere. Biological activity can transform nitrate back into the more reduced forms nitrite, nitrogen gas (N_2), and ammonium. The colorimetric determination of nitrate used in this manual measures both nitrate and nitrite, but does not quantify ammonium or nitrogen gas.

The zinc-reduction method for determination of nitrate begins with the reduction of all nitrate to nitrite using zinc metal as the reducing agent.

$$H_2O \, (l) \;+\; NO_3^- \, (aq) \;+\; Zn \, (s) \longrightarrow Zn^{+2} \, (aq) \;+\; NO_2^- \, (aq) \;+\; 2 \, OH^- \, (aq)$$

The nitrite produced then reacts with HCl and sulfanilic acid to form a diazonium salt.

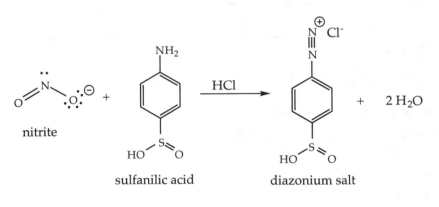

Addition of 1-naphthylamine results in the formation of a pink azo dye. The concentration of the azo dye is measured spectrophotometrically, with the intensity of the pink color proportional to the initial concentration of nitrate.

The concentration of nitrate in solution is measured using a spectrophotometer to quantify the pink azo dye formed through a series of reactions.

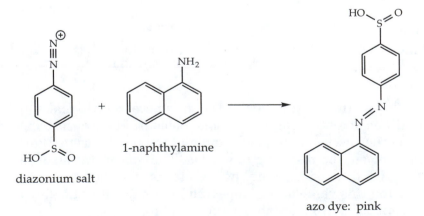

Colorimetric Method for the Determination of Phosphate in Water

In natural waters, phosphorus exists mainly in combination with oxygen in the form of phosphates. The simplest of these is **orthophosphate** (H_3PO_4, $H_2PO_4^-$, and PO_4^{-3}). There are also **condensed phosphates**, and **organically bound phosphates** found in living or decaying biological material (Figure 5-6). The colorimetric test used in this laboratory manual measures only orthophosphate.

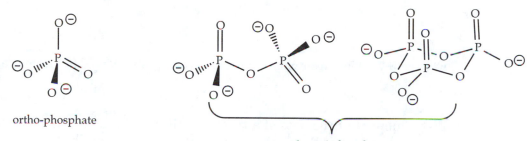

ortho-phosphate

condensed phosphates

Figure 5-6: Ortho-phosphate is the simplest type of phosphate. Condensed phosphates can consist of either linear or cyclic chains of alternating phosphorus-oxygen bonds.

The orthophosphates react with ammonium molybdate, $(NH_4)_6Mo_7O_{24} \cdot 4\,H_2O$ under acidic conditions to form molybdophosphoric acid. Reaction of this compound with ammonium metavanadate (NH_4VO_3) results in the formation of a yellow vanadomolybdophosphoric acid complex. The intensity of the absorbance of this complex at 420 nm is measured spectrophotometrically and is proportional to the original concentration of orthophosphate in the sample.

Phosphate forms a yellow complex with vanadium and molybdenum oxides.

Turbidimetric Method for the Determination of Sulfate in Water

The amount of sulfate in a water sample can be determined spectrophotometrically by employing a reaction that precipitates out very fine particles of $BaSO_4$ to make a **turbid** or cloudy solution. The chemical reaction occurring is:

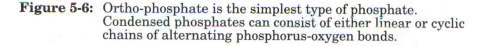

$$SO_4^{-2}\,(aq) \; + \; Ba^{+2}\,(aq) \longrightarrow BaSO_4\,(s)$$

The technique for measurement of turbidity is based on the property of light scattering by particulate matter in an aqueous solution. When a beam of light is imposed on a sample, any particles present scatter the light in all directions. The measuring device for determining turbidity is called a turbidimeter, an

Sulfate reacts with barium ions to form a fine precipitate. The amount of precipitate is determined spectrophotometrically.

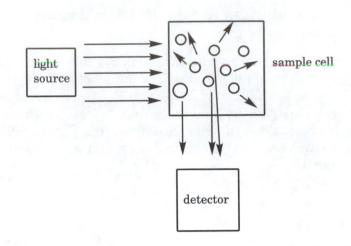

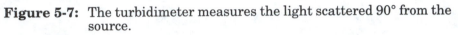

Figure 5-7: The turbidimeter measures the light scattered 90° from the source.

instrument that consists of a tungsten-filament lamp (i.e., a normal light bulb) that shines light into the sample. The instrument then detects all the light that is scattered 90° to the incident light beam (see Figure 5-7). This type of meter is also called a **nephelometer**, and turbidity is reported as **nephelometric turbidity units** or **NTU**.

Turbidity can also be measured using a spectrophotometer. The principles are the same, except that light that is not transmitted through the solution is *scattered*, not absorbed by the species in solution.

LABORATORY PROCEDURES: NITRATE, PHOSPHATE, AND SULFATE

Determining Nitrate by Zinc Reduction and Azo Dye Formation[1]

Objective: To determine the concentration of nitrate by a colorimetric method. Nitrate is reduced to nitrite with zinc and the resulting nitrite is allowed to react with sulfanilic acid and HCl to form a diazonium salt. Reaction of the diazonium salt with 1-naphthylamine produces a pink azo dye. For detailed information on the analytical method used to determine nitrate, see page 80. More information about sources of nitrate in natural waters can be found on pages 73–75.

Note All glassware used in this experiment should be acid-washed and well rinsed according to the procedures on page 25. Be sure all nitric acid is thoroughly rinsed off of the glassware.

Standardizing the Spectrophotometer

1 Using the 50 ppm (mg of NO_3^-/L) nitrate standard as the concentrated stock solution, work in groups of four to prepare the following dilute standards to calibrate the spectrophotometer. The calibration will go quickly if each person prepares one standard.

Note **Useful Tip for Calculating Volumes:** A handy formula is:
$$V_1 C_1 = V_2 C_2,$$
where V_1 is the volume of the concentrated standard of concentration C_1 that must be measured to obtain the desired volume V_2 of the diluted standard that has concentration C_2. The quantity C_2 is simply the target concentration. The volume V_2 is the volume of the volumetric flask in which the standard is prepared. Have your instructor check your calculations before beginning the preparation of the standards.

- Blank, distilled water only
- 2.0 ppm
- 5.0 ppm
- 10.0 ppm
- 15.0 ppm

2 For each of the standards, follow the procedures below.

3 Measure 50 mL of the standard (or sample) into a 250 mL Erlenmeyer flask. Add 1.0 mL of 3 M HCl and 1.0 mL sulfanilic acid solution and mix thoroughly. Add 1 mL of zinc/NaCl mixture to the sample. Note the exact time.

---

1. M. G. Ondrus, *Laboratory Experiments in Environmental Chemistry* (Wuerz Publishers, Winnipeg, 1993).

4   Mix for *exactly* 7 minutes.

⚠ ! Timing is critical for the success of the experiment!

Note

---

5   Filter the solution using a Buchner vacuum filtration setup.

6   Add 1.0 mL naphthylamine hydrochloride solution and 1.0 mL of 2 M sodium acetate solution to the filtered sample and mix well. Allow 5 minutes for development of the pink color.

7   Set the spectrophotometer to 520 nm for the analysis. Begin by zeroing the instrument with the blank, then measure the absorbance of each of the standards, noting the values in your laboratory notebook.

8   Construct a standard curve of absorbance vs. concentration (see page 78). Use the concentration of nitrate in the *original* standards, before additional reagents were added.

Sample Analysis

1   Prepare the sample in the same way the standards were prepared, using 50 mL of sample. Follow steps 3–7 above for each sample to be analyzed.

2   Using the blank solution, zero the calibrated spectrophotometer at 520 nm, then measure the absorbance of the sample. If any samples have an absorbance higher than the most concentrated standard, dilute the original sample 1:10 (see Appendix B for assistance with dilutions) and repeat the process.

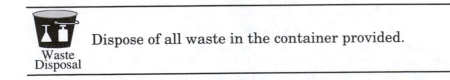   Dispose of all waste in the container provided.

Waste
Disposal

Calculations

Using your standard curve (see page 78), determine the concentration of nitrate (NO_3^-) in the water sample. Be sure to take into account any dilutions you made to bring the sample on-scale for the analysis. Convert this value to milligrams of N/L).

Determining Phosphate by Vanadomolybdophosphoric Acid Colorimetric Method[2,3]

Objective: To determine concentration of total phosphates in water by a colorimetric method. The orthophosphates react with ammonium molybdate in an acid medium to form molybdophosphoric acid. This acid then reacts with vanadium (ammonium metavanadate) to form the yellow vanadomolybdophosphoric acid complex. The intensity of the absorbance of this complex at 420 nm is proportional to the original concentration of orthophosphate in the sample. For detailed information on the analytical method used to determine phosphate, see page 81. More information about sources of phosphate in natural waters can be found on pages 73–75.

 Note All glassware used in this experiment should be acid-washed and well rinsed according to the procedures on page 25. Do not use detergent to wash glassware as it may contain phosphates.

Standardizing the Spectrophotometer

1   Using the 20 ppm (mg/L) phosphate standard as the concentrated stock solution, work in groups of four to prepare the following dilute standards to calibrate the spectrophotometer. The calibration will go quickly if each person prepares one standard.

Note **Useful Tip for Calculating Volumes:** A handy formula is:
$$V_1C_1 = V_2C_2,$$
where V_1 is the volume of the concentrated standard of concentration C_1 that must be measured to obtain the desired volume V_2 of the diluted standard that has concentration C_2. The quantity C_2 is simply the target concentration. The volume V_2 is the volume of the volumetric flask in which the standard is prepared. Have your instructor check your calculations before beginning the preparation of the standards.

- Blank, distilled water only

- 2.0 ppm, final volume of 100 mL

- 5.0 ppm, final volume of 100 mL

- 8.0 ppm, final volume of 100 mL

- 10.0 ppm, final volume of 100 mL

2   For each of the standards, use the following procedures.

2.   M. G. Ondrus, *Laboratory Experiments in Environmental Chemistry* (Wuerz Publishers, Winnipeg, 1993).

3.   A. D. Eaton, L. S. Clesceri, and A. E. Greenberg, eds., *Standard Methods for the Examination of Water and Wastewater*, 19th ed., (American Public Health Association, Washington, DC, 1995).

3 Measure 25.0 mL of the standard (or sample) into a 50-mL volumetric flask or a graduated cylinder. The solution must be acidic for the color-producing reaction to proceed properly. Determine if your solution is acidic by adding 1 drop of phenolphthalein indicator to 25.0 mL of solution. If the solution is pink, indicating pH > 8.5, add 6M HCl dropwise until the solution is colorless. Dilute the solution to 50 mL, taking care not to overshoot the mark.

4 Put 25 mL of this solution in a 50-mL volumetric flask. Add 10.0 mL vanadate-molybdate reagent with a volumetric pipet, and bring the volume to 50 mL with distilled water. Wait at least 10 minutes for the color to fully develop.

5 Zero the spectrophotometer with distilled water and measure the absorbance of the solution at 420 nm.

6 Construct a standard curve of absorbance vs. concentration (see page 78). Use the concentration of phosphate in the *original* standards, before additional reagents were added.

Sample Analysis

1 Prepare the sample in the same way the standards were prepared, using 25 mL of sample. Follow steps 3–4 above for each sample to be analyzed.

Note If your sample has color, it is important to remove the colored compounds so they do not interfere with the colorimetric determination. To remove color, shake 50 mL of the solution with about 200 mg of activated carbon for 5 minutes. Filter.

2 Using the blank solution, zero the calibrated spectrophotometer at 420 nm, then measure the absorbance of the sample. If any samples have an absorbance higher than the most concentrated standard, dilute the original sample 1:10 (see Appendix B for assistance with dilutions) and repeat the process.

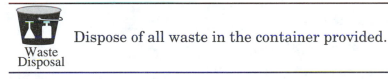

Waste Disposal Dispose of all waste in the container provided.

Calculations

Using your standard curve (see page 78), determine the concentration of PO_4^{-3} in the water sample. Be sure to take into account any dilutions you made to bring the sample on-scale for the analysis.

Determining Sulfate by a Turbidimetric Method[4,5]

Objective: To determine the amount of sulfate in water by reacting it with barium chloride to form a suspension of insoluble barium sulfate. The concentration of the barium sulfate suspension is determined by measuring absorbance at 420 nm. For detailed information on the analytical method used to determine sulfate, see page 81. More information about sources of sulfate in natural waters can be found on page 75.

Note   All glassware used in this experiment should be acid-washed and well rinsed according to the procedures on page 25.

Standardizing the Spectrophotometer

1   Using the 1000 ppm (mg/L) sulfate standard as the concentrated stock solution, work in groups of four to prepare the following dilute standards to calibrate the spectrophotometer. The calibration will go quickly if each person prepares one standard.

Note   **Useful Tip for Calculating Volumes:** A handy formula is:
$$V_1C_1 = V_2C_2,$$
where V_1 is the volume of the concentrated standard of concentration C_1 that must be measured to obtain the desired volume V_2 of the diluted standard that has concentration C_2. The quantity C_2 is simply the target concentration. The volume V_2 is the volume of the volumetric flask in which the standard is prepared. Have your instructor check your calculations before beginning the preparation of the standards.

- Blank, distilled water only

- 10.0 ppm, final volume of 100 mL

- 20.0 ppm, final volume of 100 mL

- 40.0 ppm, final volume of 100 mL

- 80.0 ppm, final volume of 100 mL

2   For each of the standards, follow the procedures below.

Note   It is important to stir the solution at a constant rate for the analysis to be consistent from sample to sample. A magnetic stirring plate is very helpful and should be used if available

4.   M. G. Ondrus, *Laboratory Experiments in Environmental Chemistry* (Wuerz Publishers, Winnipeg, 1993).

5.   A. D. Eaton, L. S. Clesceri, and A. E. Greenberg, eds., *Standard Methods for the Examination of Water and Wastewater*, 19th ed., (American Public Health Association, Washington, DC, 1995).

3 Measure 10 mL of standard (or sample) and 10 mL of distilled water into a 250-mL Erlenmeyer flask with a magnetic stir bar in it, over a magnetic stirring plate.

4 Add 6.0 mL of the specially prepared acetate buffer and stir gently (avoid splashing!) .

5 Add a pinch (about 0.1–0.2 g) of 20–30 mesh $BaCl_2$ crystals and stir the mixture for 60 ± 2 seconds. An excess of $BaCl_2$ is required for all of the sulfate to react to form $BaSO_4(s)$.

6 At the end of the stirring period, quickly transfer the solution to the spectrophotometer cell.

7 Zero the spectrophotometer at 420 nm with distilled water. Place the sample in the spectrophotometer after 5 ± 0.5 minutes and measure the absorbance.

8 Construct a standard curve of absorbance vs. concentration (see page 78). Use the concentration of sulfate in the *original* standards, before additional reagents were added.

Sample Analysis

1 Prepare the sample in the same way the standards were prepared, using 25 mL of sample. Follow steps 3–6 above for each sample to be analyzed.

2 Using the blank solution, zero the calibrated spectrophotometer at 420 nm, then measure the absorbance of the sample. If any samples have an absorbance higher than the most concentrated standard, dilute the original sample 1:10 (see Appendix B for assistance with dilutions) and repeat the process.

The solutions from this analysis may be drain disposed in most locations. Check with your instructor.

Waste
Disposal

Calculations

Using your standard curve (see page 78), determine the concentration of sulfate in the water sample. Report the sulfate concentration in ppm (mg of SO_4^{-2}/L). Be sure to take into account any dilutions you made to bring the sample on-scale for the analysis.

CHLORIDE AND FLUORIDE: SIMPLE ANIONS WITH AN IMPORTANT MESSAGE

Chloride: Important for Electrolyte Balance

Chloride ion (Cl⁻) is essential to the electrolytic balance, or balance of essential ions, in our bodies. Because there is a continuous intake and excretion of chloride from all animals, it is one of the more abundant anions found in wastewater and is a good indicator ion for pollution sources. Chloride gives water a salty taste, detectable at a level of 250 ppm if the cation present is sodium, but with calcium and magnesium as counterions, this salty taste is not detectable until the chloride concentration reaches up to 1000 ppm.

An important natural source of chloride is the dissolution of minerals and rocks by flowing water. Ocean water contains very high concentrations of chloride that originated from dissolved minerals. Anthropogenic sources of chloride in surface waters include wastewater treatment plant discharges, road salting, and manufacturing and food processing operations. Groundwater near salt water sources like the ocean may also be subject to contamination with chloride when overpumping draws salt water into the freshwater aquifer.

High chloride levels can pose a threat to crops, some animals, and freshwater aquatic plants that do not have mechanisms to excrete excess salt. High levels of chloride in a water system increase the rate of corrosion of metallic pipes.

Chloride is a good marker ion for animal, human or manufacturing wastes.

Fluoride: To Fluoridate or Not?

Fluoride occurs in minerals and is often present in a water supply naturally. However, after the discovery of a correlation between high natural levels of fluoride ion in water with a low incidence of dental cavities, many water districts decided to add fluoride to water supplies to help prevent tooth decay. Tooth decay occurs rapidly when outer tooth enamel composed of hydroxyapatite, $Ca_{10}(PO_4)_6(OH)_2$, is exposed to acids such as those present in soft drinks. Fluoride treatment replaces the reactive hydroxyl groups (OH) in the hydroxyapatite with fluoride to form the less soluble fluoroapatite, $Ca_{10}(PO_4)_6(F)_2$, which reacts much more slowly with acid. Dental fluoride treatment can be accomplished with fluoridated toothpaste and direct application of fluoride solution to the teeth, in addition to or instead of water treatment.

Excessive levels of fluoride in drinking water can cause problems. Fluoride at a concentration of 1 ppm can aid in the prevention of tooth decay, but at levels as low as 3 to 5 ppm, fluoride can cause mottling of teeth. In higher concentrations, fluoride is poisonous, causing bone cancer in rats. At very high doses it is used as a rat poison.

Fluoride is used in dental treatments to make teeth stronger and more cavity resistant.

There is some controversy over whether fluoride should be added to drinking water at all because it is poisonous at higher dosages. In this case, whether the fluoride is poisonous or beneficial is highly dependent on the dose. Thus, careful monitoring of fluoride in drinking water is necessary. If the naturally occurring fluoride in a water supply is higher than 1 ppm, it is not treated with fluoride. Naturally occurring fluoride in areas such as the southwestern United States often exceeds 4 ppm, requiring the removal of fluoride down to acceptable levels.

ELECTROCHEMICAL METHODS OF ANALYSIS

The ability of a species to gain or lose electrons serves as the basis for many analytical methods, called **electrochemical** methods. In this section, we will discuss the two electrochemical methods commonly used in the detection of chloride and fluoride in natural waters.

Potentiometric Method for Determination of Chloride in Water

One relatively inexpensive and accurate method for measuring the concentration of chloride ions is a potentiometric titration. Potentiometry uses an electrochemical cell to measure the concentration of a species based on its ability to gain or lose electrons. A specific voltage, or potential, is associated with this gain or loss of electrons and is related to the concentration of the species in solution.

Potentiometric methods utilize electrodes that measure the potential difference (voltage) as a function of concentration of a species.

The important components of a device used to monitor a potentiometric titration are:

1   A reference electrode that maintains a constant potential (voltage) throughout the titration

2   An indicator electrode that is sensitive to the concentration of one of the species in the reaction

3   A device to measure the potential (voltage) of the indicator electrode relative to the reference electrode (many pH meters can be used to measure voltage as well as pH)

4   A method for stirring the reaction system.

An electrochemical cell used to measure chloride concentration is shown in Figure 5-8.

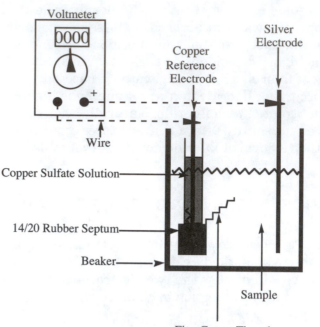

Figure 5-8: Setup for a potentiometric titration with a $Cu/CuSO_4$ reference electrode and silver indicator electrode

---

A simple electrode using copper/copper sulfate as the reference electrode and a silver wire as the working electrode measures chloride accurately.

An example of a reference electrode is a copper electrode that is isolated from the sample and connected through a cotton thread that acts as a salt bridge. Because the equilibrium between copper metal and copper(II) ions is readily established, the Cu^{+2} concentration will remain constant. This constant concentration of Cu^{+2} provides a constant voltage as a reference to the indicator electrode. For the potentiometric determination of chloride, the indicator electrode is a silver wire.

The potentiometric titration used to determine the concentration of Cl^- in an environmental sample is carried out by titrating the sample with $AgNO_3$ to form $AgCl$:

$$Ag^+(aq) + NO_3^-(aq) + Cl^-(aq) \rightleftharpoons AgCl(s) + NO_3^-(aq)$$

As Ag^+ ions are added to the sample during the titration, solid $AgCl$ precipitates out and the cell potential (or voltage) changes in response to the changing concentration of the remaining silver ions. For the specific potentiometric titration used in this experiment, the half cell reactions are:

Anode: $Cu(s) \longrightarrow Cu^{+2}(aq) + 2e^-$

Cathode: $Ag^+(aq) + e^- \longrightarrow Ag(s) + 2e^-$

Net: $2Ag^+(aq) + Cu(s) \longrightarrow 2Ag(s) + Cu^{+2}(aq)$

Because [Ag⁺] and [Cl⁻] are related by their solubility equilibrium,

$$AgCl\ (s) \rightleftharpoons Ag^+\ (aq)\ +\ Cl^-\ (aq) \qquad K_{sp} = 1.8 \times 10^{-10}$$

the cell indirectly measures the concentration of Cl⁻ by permitting the determination of how much silver is required to precipitate out the chloride present.

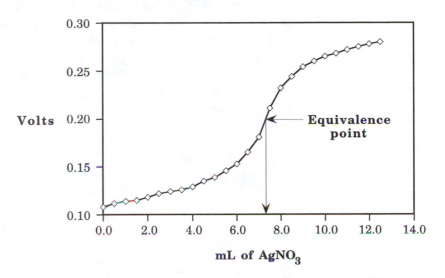

Figure 5-9: Potentiometric titration curve of voltage vs. milliliters of AgNO₃ added for a solution containing 20 ppm chloride. The equivalence point is the point where the slope of the curve is steepest.

The equivalence point of the titration is determined by plotting cell potential vs. volume of AgNO₃ added (see Figure 5-9) and noting the point on the curve with the steepest slope. The shape of the titration curve is dependent on the log of the concentration of silver ion squared relative to the log of the constant copper ion concentration. Before the equivalence point, almost all of the Ag⁺ added precipitates as AgCl. At the equivalence point, the concentration of Ag⁺ is equal to the concentration of Cl⁻, as dictated by the K_{sp} given above. The volume of AgNO₃ solution added can be read off of the graph at this point and the moles of Ag⁺ required to reach the equivalence point can be calculated. Because chloride ion reacts in a 1:1 stoichiometry with silver ion, the moles of silver calculated are equal to the moles of chloride ion initially contained in the sample.

The number of moles of chloride in a sample is equal to the number of moles of silver ion added to reach the equivalence point.

Ion-Selective Electrode Determination of Fluoride in Water

A useful technique for the analysis of specific ions in water involves the use of **ion-selective electrodes** or **ISEs**. Ion-selective electrodes are designed to respond selectively to a given ion and use **potentiometry** to provide a quantitative measurement of concentration. The pH meter (see page 56) is an ion-selective electrode that is specific for hydronium ions (H_3O^+). There are a variety of ion-selective electrodes commercially available, including those for fluoride, sodium, ammonium, cyanide, and copper.

Specific determination of fluoride is routinely carried out using an ISE (Figure 5-10). The fluoride ISE works by using an electrical circuit that contains the ISE, an external reference electrode, and a voltmeter that gives a measure of the electrical potential in the circuit. The fluoride electrode uses a chemically modified (doped) lanthanum fluoride crystal that is in contact with both the test solution and an internal reference solution. Differences in fluoride concentration across the crystal generate a potential which is measured by the voltmeter.

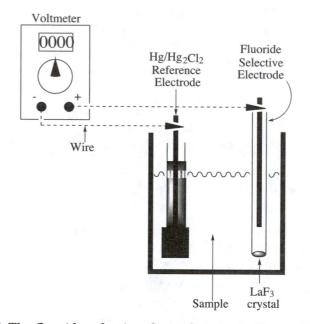

Figure 5-10: The fluoride-selective electrode is typically used in conjunction with a saturated calomel (Hg/Hg_2Cl_2) electrode.

A standard calomel electrode (Hg/Hg_2Cl_2) is used as the reference electrode. The cell is represented as:

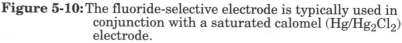

Ag | AgCl, 0.3 M Cl⁻, 0.001 M F⁻ | LaF$_3$ | test solution | reference electrode

The electrochemical potential of the cell is defined by the Nernst equation, which, for most freshwater samples, simplifies to:

$$E_{cell} = \text{constant} -0.0592 \log [F^-]$$

The Nernst equation describes the voltage of an electrochemical cell as a function of concentration of a particular ion.

where E_{cell} is in millivolts and the constant is a characteristic of the particular electrochemical cell. Before the readings can be considered to be accurate, the meter must first be standardized by placing the electrodes into solutions of known F^- ion concentration and determining a linear relationship between log $[F^-]$ and millivolts. If high concentrations of other anions are present in the sample, correction factors are necessary to obtain reliable data.

During the analysis, a buffer solution is added to maintain the pH between 5.2 and 5.7. This serves both to reduce the concentration of interfering hydroxide ions, as well as ensure that most of the fluoride exists as F^-, not as the weak acid HF (pK_a = 3.17). The buffer also contains a complexing agent, either cyclohexylenediaminetetraacetic acid or citric acid, that competes for ions such as Al^{+3} and Fe^{+2} that typically bind fluoride strongly. This reagent serves to free the fluoride ions so the electrode effectively measures F^- concentration.

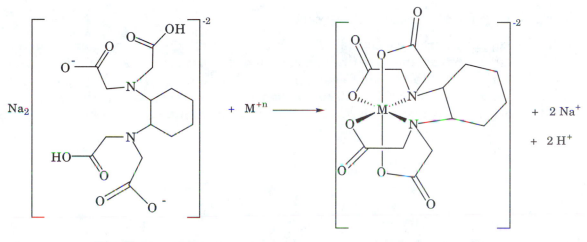

Disodium cyclohexylenediaminetetraacetic acid Na$_2$ (CDTA)

$[M(CDTA)]^{-2}$ **complex**

Because the electrode responds differently to solutions of different ionic strength, the fluoride measurement is best carried out under conditions of high ionic strength. This minimizes the change in electrode response as a function of the concentration of other ions in the solution.

LABORATORY PROCEDURES: CHLORIDE AND FLUORIDE

Determining Chloride by Potentiometric Titration[6,7,8]

Objective: To determine chloride concentration by a precipitation titration with silver nitrate. The endpoint of the titration is determined potentiometrically. See page 91 for more information on potentiometric methods and page 90 for more information on sources of chloride ions in natural waters.

⚠ **Note**   All glassware used in this experiment should be acid-washed and well rinsed according to the procedures on page 25.

Preparing the Copper Reference Electrode[9]

1   To make the copper reference electrode, first obtain a 6-inch long 8 mm outer-diameter glass tube. Make sure the ends are fire polished to minimize the chances of injury. On one end of the glass tube, place the fine cotton thread so that half of the length of the thread is inside the tube. Place a 14/20 rubber septum on the same end as the thread so that half of the thread is inside the tube, and the other half outside.

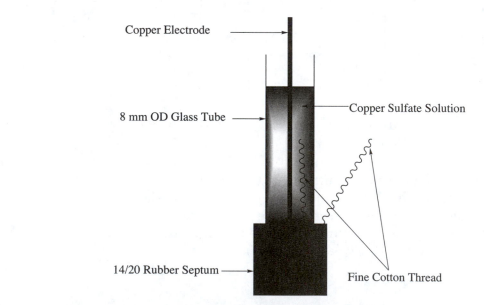

2   Cut a piece of copper wire about the same length as the glass tube. This will be the copper electrode. Clean the copper wire with steel wool until the copper is shiny. Rinse well. Attach an alligator clip with a wire lead to one end of the copper wires and insert the other end into the open end of the glass tube.

---

6.   G. Lisensky and K. Reynolds, *J. Chem. Ed.*, 1991, v. 68, 334–335.

7.   D. C. Harris, *Quantitative Analysis* (W. H. Freeman & Co., New York, 1995).

8.   A. D. Eaton, L. S. Clesceri, and A. E. Greenberg, eds., *Standard Methods for the Examination of Water and Wastewater*, 19th ed., (American Public Health Association, Washington, DC, 1995).

9.   R. Ramette, *Chemical Equilibrium* (Addison-Wesley, Reading, MA, 1981), p. 649.

3 Fill the tube half to three-quarters full with 0.1 M $CuSO_4$. Place the electrode upright in a 150-mL beaker.

Preparing the Silver Electrode

1 It is important to remove all the tarnish from the silver electrode so that its behavior can be predicted by the Nernst equation. A quick and simple way to clean the electrode is by placing it in concentrated nitric acid for a few minutes. Coil the electrode so that it rests flat on the bottom of a 100 mL beaker and add enough 6 M nitric acid to cover the electrode (this should be only a few mL). Allow the electrode to soak for a few minutes, but do not leave it too long as the acid will dissolve the silver. Dispose of the acid in the appropriate waste container and rinse the electrode well.

 Caution Nitric acid will burn if spilled on the skin or in the eyes. It is imperative that goggles be worn at all times. If you spill acid on your skin, flush the affected area with copious amounts of water. Know where the closest eyewash, shower, and sink are located. If you spill acid on the lab bench, put a *small* amount of baking soda on the spill to neutralize the acid and then wipe up the excess baking soda. DO NOT leave unlabeled containers of acid on your laboratory benchtop.

2 Attach an alligator clip with a wire lead to one end of the silver electrode and place the other end into the 150-mL beaker beside the copper reference electrode. Connect the wire from the silver electrode to the positive terminal of the voltmeter.

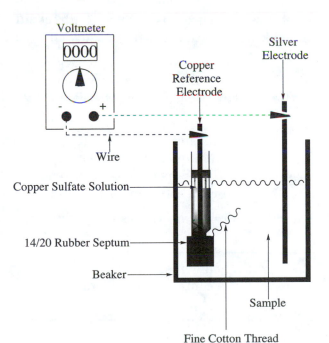

Sample Analysis

1   Acid wash a buret (see procedures on page 25) and rinse well with distilled water. Rinse buret with a small amount of 0.004 M $AgNO_3$ solution, taking care to coat the walls with it. This avoids dilution of the standard titration solution.

> Silver nitrate irritates and discolors the skin. It is also an eye irritant.

Caution

2   Fill the buret with 0.004 M $AgNO_3$ solution. Open the stopcock and let the silver nitrate flow to remove air bubbles from the tip of the buret, then refill the buret until it contains at least 25 mL of 0.004 M $AgNO_3$. Measure 50.0 mL of sample with a graduated cylinder and pour it into a 150-mL beaker.

> A 50.0-mL sample provides a good endpoint for samples with chloride concentrations between 5 and 20 ppm. If you suspect your sample has a chloride concentration less than 5 ppm, use 100 mL of sample for the titration. If you suspect your sample has a chloride concentration between 20–100 ppm, use 25 mL of sample and 25 mL of distilled water for the titration. For seawater samples, where chloride concentrations are extremely high, the concentration of $AgNO_3$ will need to be ~0.4 M.

Note

3   Turn on voltmeter and be sure it is in the DC (direct current) mode. Adjust the meter to read in **Volts**, making sure that it will measure in the range from 0–2 V.

4   Before adding the titrant, record the voltage. Begin adding titrant in 0.5-mL aliquots. Stir and let the solution equilibrate, then record the voltage to the nearest millivolt. As you approach the equivalence point, the change in the voltage with each aliquot added becomes more dramatic.

5   Keep titrating well past the endpoint, continuing the titration until changes in the voltage on addition of each aliquot are on the order of 2 to 3 mV. Add an additional 5 mL at the end to verify that the endpoint has been reached. A typical volume of titrant required for surface waters is between 10 and 50 mL.

6   Perform two additional replicate analyses to ensure the accuracy of the analysis.

> Silver is a heavy metal and cannot be drain disposed. Dispose of all waste in the container provided.

Waste Disposal

Calculations

1 Plot the data as volts vs. mL of $AgNO_3$. From this curve, the equivalence point can be determined by noting the inflection point (the point where the slope is steepest) for the curve. Read the volume of $AgNO_3$ at this point and use the concentration to determine the moles of $AgNO_3$ required to reach the equivalence point. The concentration of Cl^- ions in the sample can then be calculated by knowing the stoichiometry of the reaction of Ag^+ with Cl^-.

2 Report the results in both moles of Cl^-/liter and in milligrams of Cl^-/L. Be sure to take into account any dilutions you might have made on your sample.

Determining Fluoride by Ion-Selective Electrode[10]

Objective: To determine the concentration of fluoride ion in water using a fluoride-selective electrode. For details on how a fluoride-selective electrode operates, see page 93. More information about fluoride in natural waters can be found on page 90.

Note

The fluoride and reference electrodes are extremely delicate and costly instruments, and are especially fragile on the bottom. You must take extreme caution not to handle the bottom of the electrodes, nor bump them against the beaker. Also, make sure that the electrode is above the stir bar when you are testing your solution.

Note

All glassware used in this experiment should be acid-washed and well rinsed according to the procedures on page 25.

Calibrating the Fluoride Electrode

1   Using the 100-ppm (mg/L) fluoride standard as the concentrated stock solution, work in groups of four to prepare the following dilute standards to calibrate the ion-selective electrode. The calibration will go quickly if each person prepares one standard.

> **Note**
>
> **Useful Tip for Calculating Volumes:** A handy formula is:
> $$V_1 C_1 = V_2 C_2,$$
> where V_1 is the volume of the concentrated standard of concentration C_1 that must be measured to obtain the desired volume V_2 of the diluted standard that has concentration C_2. The quantity C_2 is simply the target concentration. The volume V_2 is the volume of the volumetric flask in which the standard is prepared. Have your instructor check your calculations before beginning the preparation of the standards.

- Blank, distilled water only

- 1.0 ppm, final volume of 100 mL

- 2.0 ppm, final volume of 100 mL

- 5.0 ppm, final volume of 100 mL

- 7.0 ppm, final volume of 100 mL

2   For each standard, follow the procedures below. Begin the calibration procedure with the lowest concentration standard and end with the highest concentration standard.

10. A. D. Eaton, L. S. Clesceri, and A. E. Greenberg, eds., *Standard Methods for the Examination of Water and Wastewater*, 19th ed., (American Public Health Association, Washington, DC, 1995).

3 Add 10 mL of fluoride buffer to 10.0 mL of standard or sample. Use volumetric pipets to transfer samples. Be sure all solutions are the same temperature.

4 Adjust the calibration control on the voltmeter so that the 1.0 mg F⁻/L standard reads 100 mV in the expanded scale position.

5 Place the electrode in each standard solution, beginning with the lowest concentration standard. Equilibrate for 3 minutes and read the potential from the voltmeter.

6 Construct a standard curve of **mV** vs. **log (mg F⁻/L)** for the standard solutions. Use the concentration of fluoride in the *original* standards, before additional reagents were added.

Sample Analysis

1 Prepare the sample in the same way the standards were prepared, using 10.0 mL of sample.

2 Repeat steps 3–5 for each sample to be analyzed. If any samples have a millivolt reading higher than the most concentrated standard, dilute the sample 1:10 (see Appendix B for assistance with dilutions) and repeat the test, beginning with step 3.

Waste
Disposal

The solutions from this analysis may be drain disposed in most locations. Check with your instructor.

Calculations

Using your standard curve, determine the concentration of F⁻ in the sample in milligrams per liter. Be sure to take into account any dilutions.

Chapter 6

Ion Chromatography

Ion chromatography permits simultaneous determination of several ions.

It is often useful to know what types of ions are present in a water sample and the concentrations of *specific* ions. For example, you may be interested in knowing specifically how much chloride (Cl^-) is leaking into an underground well from a nearby ocean. Or perhaps you are interested in the amount of nitrate (NO_3^-) from the fertilizer that is running off of a farmer's field into a river. Or maybe you would like to know how much sulfate (SO_4^{-2}) from coal-burning power plant emissions ends up in rainwater. **Ion chromatography (IC)** is a technique used to measure trace concentrations of a variety of different **anions**, or *negatively* charged ions such as fluoride, chloride, nitrate, sulfate, and phosphate.

Ion chromatography is a modern method that enables the chemist to determine not only how much of a given ion is present but also to determine specifically which ions are present in solution. It works by first separating the ions from each other and then using a conductivity detector to measure the concentrations of the individual ions. There are many important chemical principles contained in the theory and practice of ion chromatography. Let's look at them in some detail.

SEPARATION OF IONS

Chromatography is defined as the process by which a mixture of substances is separated into its component parts. There are many different types of chromatography; ion chromatography is specifically for the separation of a mixture of ions. How is this accomplished?

All chromatographic methods of separation rely on several fundamental principles:

1   There is a **mobile phase** that carries the mixture of compounds through the separation column.

2   There is a **stationary phase** in the column onto which different compounds adsorb (stick) for different periods of time. Some compounds are "stickier" than others, requiring more time to wash them off or **elute** them from the column.

3   There is a **force** that pushes the components of the mixture through the column. This force can be gravity, capillary action, a high pressure pump, or gas pressure.

In practice, ion chromatography is carried out by injecting an aqueous solution containing a mixture of ions is onto a **chromatography column**, a tube 4–8 inches long and approximately 1/2 inch in diameter that contains an **ion-exchange resin** (the *stationary phase*) that binds specifically to either cations or anions. The ions are washed through the column with an **eluant** (the *mobile phase*) using a high pressure pump (the *force*). For a given experimental "run," either cations or anions can be measured, but not both simultaneously. For an example of how the resin works, consider an **anion-exchange resin**. The resin itself is made of an unreactive organic polymer (similar to a plastic) that serves as a binding site for positively charged groups called **quaternary ammonium salts**.

These salts are similar in structure to ammonium ions (NH_4^+), but they are chemically bound to the resin surface (see Figure 6-1) and are not free to move. The counterions are free-moving bicarbonate ions.

As the sample containing ionic substances passes over the resin, an equilibrium is established whereby the anions in the sample replace the bicarbonate ions that are bound to the positively charged sites on the resin.

Separation of ions is carried out using an ion-exchange resin. Ions bind to the resin, with certain ions moving faster than others through the column.

$$[\text{Resin}]^+\text{-HCO}_3^- \ + \ \text{Cl}^- \ \rightleftharpoons \ [\text{Resin}]^+\text{-Cl}^- \ + \ \text{HCO}_3^-$$

The sample anions *exchange* for the bicarbonate ions, hence the name **ion-exchange**. The sample is washed through the column or **eluted**, with a solution of sodium bicarbonate. This washing solution is called the **eluant**. There is a constant competition between the bicarbonate ions and the sample anions for binding to the positively charged sites on the resin. Because there are so many more of the bicarbonate ions present from the continuing influx of eluant, eventually all of the sample anions are displaced again by bicarbonate ions and washed off of the column. A typical chromatographic "run" on an ion chromatograph may take anywhere from three to ten minutes.

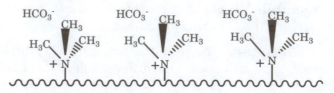

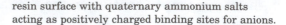

resin surface with quaternary ammonium salts
acting as positively charged binding sites for anions.

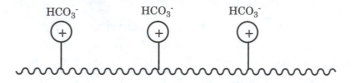

a simplified schematic of the resin surface with its
associated bicarbonate ions.

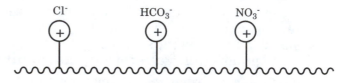

Addition of a sample containing nitrate and chloride ions
results in the displacement of some of the bicarbonate
ions on the resin surface.

Figure 6-1: An anion exchange resin works by exchanging bicarbonate
ions for sample anions.

Each ion has a characteristic retention time, or time that it takes for the ion to travel through the column and reach the detector.

Separation of sample anions occurs because different anions bind to the resin with different strengths, and therefore for different periods of time, depending on the anion. The relative affinities of the different anions for the resin is related to their charge and size as well as to their ability to interact with the positively charged site on the resin. The amount of time required for each ion to elute from the column is called the **retention time** of the ion. Figure 6-2 shows a schematic of an ion chromatograph.

IDENTIFICATION OF IONS

After the sample exits the separation column, it passes through a **suppressor** where it is treated with a source of acid to transform the bicarbonate ions in the eluant into the neutral, nonionic compound carbonic acid, H_2CO_3.

$$HCO_3^- + H_3O^+ \longrightarrow H_2CO_3$$

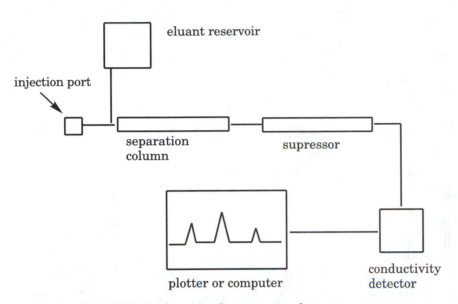

Figure 6-2: Schematic of an ion chromatograph.

The purpose of this step is to transform the eluant from an ionic to a neutral species. This is an important step because the method used to detect the ions at the end of the column is a **conductivity detector**, which measures the amount of current carried by the solution as it exits the column. As the sample ions flow through the detector, the current increases. A plot of conductivity versus time is generated and when sample ions pass through the detector, a peak is formed on the graph (see Figure 6-3). The *area* under the peak is proportional to the amount of that particular ion in solution.

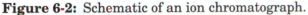

The area of a peak on an ion chromatogram is proportional to the concentration of the species in solution.

In order to determine exactly how much of each ion is present and to determine which peak is correlated to a specific ion, the instrument must first be **standardized** by analyzing a sample with *known* concentrations of the ions being analyzed and correlating the known concentrations with the *area* of the peak produced by the detector. This operation is usually carried out by a computerized data acquisition system.

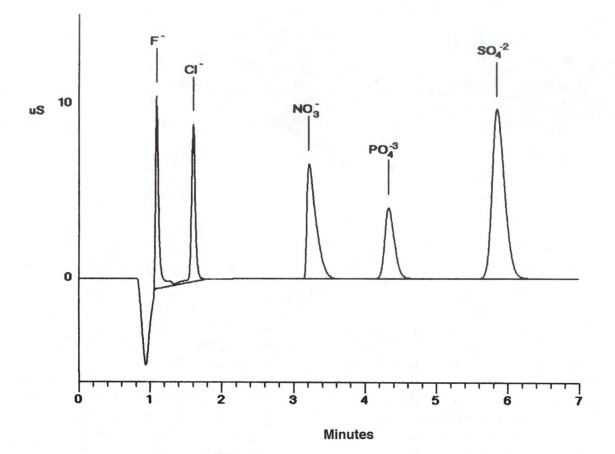

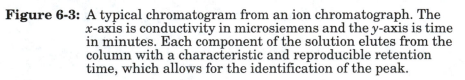

Figure 6-3: A typical chromatogram from an ion chromatograph. The *x*-axis is conductivity in microsiemens and the *y*-axis is time in minutes. Each component of the solution elutes from the column with a characteristic and reproducible retention time, which allows for the identification of the peak.

LABORATORY PROCEDURES: ION CHROMATOGRAPHY

Objective: To identify the anions of strong acids, fluoride (F^-), chloride (Cl^-), nitrate (NO_3^-) as N, phosphate (PO_4^{-3}), and sulfate (SO_4^{-2}) using ion chromatography. For detailed information on ion chromatography, see pages 102–106. More information about sources of these anions in natural waters can be found on pages 73–91 and in Appendix A.

 A *filtered* sample (0.45 µm pore size filters) should be used for ion chromatography.

Note

 All glassware used in this experiment should be acid-washed and well rinsed according to the procedures on page 25.

Note

Standardization of the Instrument

1   Using the detailed instructions specific to your instrument, inject the standards into the instrument one at a time to obtain the areas of each peak at different concentrations of the ion.

2   Plot concentration (*x*-value) versus peak area (*y*-value) for each component in the standard mixture and obtain the slope of the line by doing a least squares linear regression fit of the data points to obtain the best straight line. The computer attached to your instrument will likely perform this analysis for you.

Running the Sample

Following the detailed instructions specific to your instrument, inject the filtered sample and use the standard curve for each ion to obtain concentrations of fluoride (F^-), chloride (Cl^-), nitrate (NO_3^-) as N, phosphate (PO_4^{-3}), and sulfate (SO_4^{-2} in mg/L. If the anions are too concentrated and the peaks go off-scale, dilute the sample to an appropriate level, using pipets and a volumetric flask (see instructions in Appendix B). Be sure to note the dilution factor in your notebook and take it into account when you evaluate your data.

All solutions from this analysis may be drain disposed.

Waste Disposal

Calculations

1   Determine the concentration of each ion in moles per liter and parts per million.

2   Determine the total moles of negative charge in the solution by taking the concentration of each ion in moles per liter and multiplying it times the charge on the ion. Sum these numbers to obtain the negative charge in the sample due to the anions measured by the ion chromatograph. Combine this result with the results of the alkalinity determination to find the total amount of negative charge in the water sample.

Chapter 7

Cations in Natural Waters

The most common inorganic cations found in natural waters are calcium (Ca^{+2}), magnesium (Mg^{+2}), sodium (Na^+), and potassium (K^+). Iron (Fe^{+2}) occurs in lower concentrations, but has a significant effect on water quality because it stains fixtures and makes water taste bad. Although there are lower concentrations of other cationic components in natural waters, these common cations are in high enough concentration to be easily monitored and can readily be used as indicators of sources of pollution.

CALCIUM AND MAGNESIUM: WATER HARDNESS

Hardness and alkalinity are related water quality parameters.

The ions that contribute to **water hardness** are Ca^{+2} and Mg^{+2}, and to a lesser extent Ba^{+2} and Sr^{+2}. These ions are most often found as the counterions that accompany the alkalinity ions carbonate, bicarbonate, and hydroxide, thus the alkalinity of a water sample can be a reasonable guide to water hardness as well (page 62).

If you live in a city or town with hard water, you probably know that you have to work quite hard to get a good lather out of your bar of soap. The reason for this is that the hardness ions, predominantly Ca^{+2} and Mg^{+2}, react with soap to form an insoluble precipitate that is the calcium or magnesium salt of the soap molecule. This precipitation reaction removes the soap from solution and prevents it from solubilizing dirt.

$$Ca^{+2} \text{(aq)} \; + \; H_3C\text{-}(CH_2)_{10}\text{-}COO^- \text{(aq)} \longrightarrow [H_3C\text{-}(CH_2)_{10}\text{-}COO^-]_2 \, Ca^{+2} \text{(s)}$$

hardness ion   **soap**   **insoluble precipitate (soap scum)**

The advantage of detergents over soaps is that detergents do not form precipitates with calcium and magnesium.

Although hard water is merely an annoyance when you are trying to shampoo your hair, it can become a major problem for hot water heaters or boilers used in homes or industry. Heating water that contains calcium or magnesium bicarbonate results in the precipitation of $CaCO_3$ or $MgCO_3$. Inside a hot water heater or a boiler, these carbonate compounds precipitate out on the heater coils and in the pipes. This "boiler scale" reduces the efficiency of heater coils and blocks the flow of water through pipes. Boiler scale may also form as a white precipitate on a tea kettle used for boiling hard water. Thus, for many home and industrial applications it is often desirable to soften the water to remove as many of the hardness ions as possible.

The source of most calcium and magnesium in water supplies is natural. Minerals such as gypsum ($CaSO_4 \cdot 2H_2O$), limestone containing calcite ($CaCO_3$), and dolomite ($CaMg(CO_3)_2$) all contribute to water hardness.

Hard water damages pipes and fixtures and reduces the effectiveness of soaps.

DETERMINING WATER HARDNESS

As water hardness increases, the sudsing ability of soaps is greatly decreased. One way to measure hardness is to measure how long an inch of soap suds lasts. We will use a more accurate method to determine the concentrations of the alkaline earth cations. The final result will be reported as mg of $CaCO_3$ per liter.

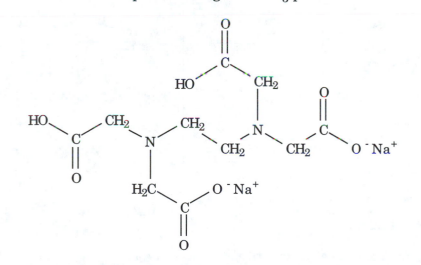

Figure 7-1: Ethylenediaminetetraacetic acid is usually used as the disodium salt. Note the large number of oxygen and nitrogen atoms in the molecule that will bind to metal ions such as calcium and magnesium.

Water hardness is measured as the sum of calcium and magnesium ions. One technique for carrying out this measurement is by titration using ethylenediaminetetraacetic acid (abbreviated as EDTA), shown in Figure 7-1. EDTA contains bonding sites

EDTA is an organic molecule that binds to calcium and magnesium to form a stable complex.

(electron pairs) that attract positive ions such as calcium and magnesium. When these ions bond to the EDTA a complex is formed called a chelate (see Figure 7-2). Metal chelates are very important in nature and a common means of transport of metals in the environment. Chelates are very soluble and keep metals in solution when they would normally precipitate and be relatively immobile. Other examples of chelate complexes in living systems are magnesium in chlorophyll and iron in hemoglobin. You will be using the disodium salt of EDTA for the analysis. This dianion, $[EDTA-H_2]^{-2}$ reacts with Ca^{+2} and/or Mg^{+2} (or any metal ion, M, with a +2 charge or greater) according to the following reaction:

$$[EDTA-H_2]^{-2} + M^{+2} \longrightarrow [EDTA-M]^{-2} + 2 H^+$$

An indicator dye, Eriochrome Black T, is added that also forms a complex with calcium and magnesium, albeit a weaker complex than the one formed with EDTA. As the EDTA is added the dye begins to complex the calcium and magnesium ions. At the endpoint, the EDTA begins to take Ca^{+2} and Mg^{+2} away from the weaker dye complex, causing a dramatic change in the color of the dye. Dye bound to calcium or magnesium is wine red. Free (unbound) dye is blue. The titration requires a buffer, because hydrogen ion is released when the EDTA binds the metal ions and dye colors are very sensitive to pH.

A color change from raspberry red to blue signals the endpoint of the EDTA titration for calcium.

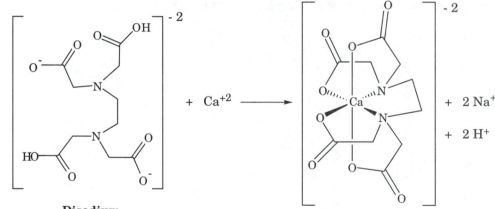

Na₂

**Disodium
ethylenediaminetetraacetic acid
Na₂ (EDTA)**

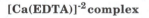
[Ca(EDTA)]⁻² complex

Figure 7-2: EDTA reacts with metal ions to form a tightly bound chelate complex that is often much more water soluble than the free metal ion.

An alternative method for determination of the concentration of calcium and magnesium in water is atomic absorption spectrophotometry. This method requires more expensive instrumentation, but provides more accurate results. This instrumental technique is discussed in more detail in a following section of this laboratory manual.

LABORATORY PROCEDURES: WATER HARDNESS

Water hardness refers to the quantity of alkaline earth cations dissolved, predominantly Ca^{+2} and Mg^{+2}, and to a lesser extent Ba^{+2} and Sr^{+2}. These ions are most often found as the counterions that accompany the alkalinity ions carbonate, bicarbonate, and hydroxide. Thus the alkalinity of the water sample can be a reasonable guide to water hardness as well. In this experiment, you will determine the water hardness by EDTA titration.

Objective: To determine the concentration of the water hardness ions Mg^{+2} and Ca^{+2} by titration with EDTA. For more information about the hardness ions, see p. 108 and for more information about the EDTA titration, see page 109.

Note

All glassware used in this experiment should be acid-washed and well rinsed according to the procedures on page 25.

Note

Work in groups of three. Each group member should do the analysis and the group should compare the results obtained. If the three results are not within ±5% of each other, repeat the analysis.

Determining Total Hardness (Ca^{+2} and Mg^{+2}) by EDTA Titration

1   Pipet 5.0 mL of the water sample into a 250-mL Erlenmeyer flask. Add:

- 50.0 mL of distilled water

- 3.0 mL of pH 10 buffer

- 6 drops of Eriochrome Black T indicator

- 1 mL of a 1:1 Mg:EDTA solution (NOTE: Your pH 10 buffer may already contain this Mg:EDTA complex. Check with your instructor.)

Note

The volume of water sample required for an optimum titration may be different than 5 mL, depending on the hardness of the water sample. If the titration requires you to refill your buret, you should use less sample. If only a few mL of EDTA titrant are required to reach the endpoint, redo the titration using more sample. Be sure to note the actual volume in your laboratory notebook and use that value in your calculations.

2   Fill your buret with the EDTA. Be sure to record the starting volume of the EDTA in the buret to at least two decimal places. This can be achieved by estimating the distance from the lines on the buret.

3   Begin to titrate, swirling the flask and using water from your wash bottle to wash down the sides of the flask periodically. The solution will change from a raspberry red color (if

significant Ca^{+2} and/or Mg^{+2} are present) to blue-violet at the endpoint. If you overshoot the endpoint, the solution will be blue with no hint of purple. If you miss the endpoint, you can add additional sample (be sure to keep track of exactly how much you add) and try approaching the endpoint again. An ideal titration will use between 10 and 20 mL of EDTA.

4   Keep the flask to use for comparison with the samples done by the other two members of your group. This will allow for direct comparisons of the endpoint colors, which should be the same. The three titrations should differ by less than 5%. If this is not the case, repeat the titration.

Determining Ca^{+2} Content by EDTA Titration

The following procedure will allow you to determine the calcium separately. This is achieved by precipitating the Mg^{+2} cations as $Mg(OH)_2$.

1   Pipet 10.0 mL of the water sample into a 250-mL Erlenmeyer flask. Add:

- 50.0 mL of distilled water

- 30 drops of a 50% NaOH solution

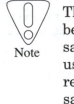

Note

The volume of water sample required for an optimum titration may be different than 10 mL, depending on the hardness of the water sample. If the titration requires you to refill your buret, you should use less sample. If only a few milliliters of EDTA titrant are required to reach the endpoint, redo the titration using more sample. Be sure to note the actual volume in your laboratory notebook and use that value in your calculations.

2   Allow 5 minutes for the magnesium to precipitate. You may not see any precipitate, but precipitation will be occurring anyway.

3   Meanwhile, weigh out ~0.1 g of hydroxynaphthol blue indicator. This indicator's color is more stable than Eriochrome Black T at a higher pH. Add the indicator solid to your sample, swirl to dissolve the powder, and titrate the sample to a blue-violet endpoint.

4   After you reach the indicator endpoint, let your sample sit for 5 minutes. This allows any $Ca(OH)_2$ that may have been formed to redissolve. Add a few more drops of EDTA if necessary and record this volume as the endpoint for the titration. The sample may change color again upon standing, which is largely due to the instability of the indicator. Just ignore further color changes. If the results of the three titrations do not agree within 5%, repeat the titration.

Waste Disposal

Dispose of the waste in the container provided. Be sure to include any extra EDTA solution from your buret.

Calculations

1   Calculate the total hardness in the sample as milligrams of $CaCO_3$ per liter (remember, this unit also includes the Mg^{+2}). You will need to know the stoichiometry of the chelation reaction, as shown on page 110.

2   Calculate the concentration of Ca^{+2} and Mg^{+2} in the sample in moles per liter and in milligrams per liter.

SODIUM AND POTASSIUM

Along with the hardness ions calcium and magnesium, sodium (Na^+) and potassium (K^+) ions are in fairly large concentration in natural waters. This is because of their high natural abundance and high water solubility. You are probably familiar with salty ocean water, which has a concentration of 10,000 mg/L of sodium and about 100 mg/L of potassium. With the oceans occupying approximately 70% of the surface area of the earth, you can well imagine that there is no shortage of sodium and potassium!

Sodium and potassium are essential elements, but too much in a water supply makes the water taste salty or bitter.

Both sodium and potassium are essential to the proper function of biological systems. In higher animals, these two ions play a major role in the transmission of nerve impulses and the intracellular neutralization of charged biomolecules such as DNA and RNA. Because of the much higher abundance of sodium over potassium in natural waters, biological systems have developed mechanisms for concentrating potassium and excluding sodium from cells to obtain the necessary amounts of both ions. In spite of this ability to select which ions are taken up, is still possible to overload organisms with sodium. The result is usually death.

Natural sources of sodium are deposits of dried ocean beds containing evaporite minerals such as halite ($NaCl$) and mirabilite (Na_2SO_4). Potassium is less abundant and is contained in a variety of minerals, including feldspars, mica, and clays.

High sodium levels in water are undesirable for a drinking water supply, both because of taste and because at high enough concentrations, salt water is an emetic, inducing nausea and vomiting. Hence the remark of a sailor who had run out of fresh water on his ship: "Water water everywhere, nor any drop to drink!"

In areas where there are no naturally occurring sodium-containing minerals, a high level of sodium in water is usually a good indicator that human or animal activities have polluted the water supply. Sewage treatment plants, landfills, the salting of roads in winter, and agricultural activities all contribute sodium to surface waters. Near coastlines, the pumping of groundwater often draws salt water into the wells through porous rock connecting the water table and the ocean, leading to high sodium levels. In delta areas where rivers meet the oceans, removal of freshwater from rivers for agricultural and municipal water supplies can often move the zone of salinity up the delta far enough to threaten fish and other aquatic organisms that cannot tolerate high salinity.

© Gilbert van Ryckevorsel

Determining Total Cation Content

The best method for determining the concentration of potassium and sodium in natural waters is atomic absorption spectrophotometry (see the following section, pages 120-125). However, **ion-exchange chromatography** is a cheaper method that can be employed to analyze for total cation content of a sample. Although the ion exchange method does not distinguish between the different cations, it can be used in combination with the water hardness analysis to determine the sum of sodium and potassium in a sample.

Synthetic and natural ion exchange resins are of considerable importance in various chemical processes. An ion exchange resin is a substance that has the capacity to exchange ions with those found in a solution. The method outlined in the procedures makes use of a synthetic cation exchange resin, Dowex 50 resin, which consists of randomly cross-linked polystyrene sulfonic acid. The organic framework of the resin bead is schematically shown in Figure 7-3.

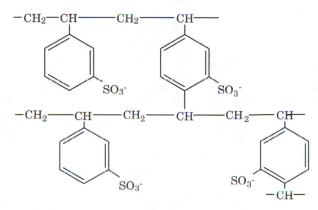

Figure 7-3: Dowex 50 resin consists of an organic polymer backbone with pendant sulfonate groups to bind positive ions.

The strings of carbon–carbon bonds create cross-linking bonds that hold the resin bead together although an appreciable amount of water can still enter the resin bead. Thus, although the resin appears to be solid, it contains many holes and passageways filled with water. For each negative sulfonate ($-SO_3^-$) group there must be a cation in order to maintain charge neutrality. These cations are omitted from the framework pictured in the figure in order to emphasize that the positive ions present in the resin are not necessarily attached to any particular $-SO_3^-$ group, just as sodium ions are not attached to SO_4^{-2} ions in a solution of Na_2SO_4. Although the resin is a solid, it resembles a concentrated aqueous solution, one in which the anions are fixed and cannot move while the cations are free to move and can even leave the resin bed entirely if replaced by another cation.

An ion exchange resin has negatively charged sites on its surface that bind to positively charged cations.

The Chemistry of Ion Exchange

Let's look into the chemistry of ion exchange in more detail. If the Dowex 50 resin, containing for example hydrogen ions, is put in contact with NaCl solution, some of the hydrogen ions will diffuse out of the resin, and a corresponding number of sodium ions will diffuse into the resin. This process will continue until equilibrium is reached. Because the cations in the resin are not strongly attached to the $-SO_3^-$ groups, the resin exhibits only a mild preference for one cation over another, as long as they have the same charge.

The equilibration process can most simply be thought of as the replacement of one ion. Thus, for the foregoing case we would write the reaction that is occurring as:

$$Na^+ (aq) + H^+ (r) \rightleftharpoons Na^+ (r) + H^+ (aq)$$

where r and aq indicate the resin and aqueous phases, respectively. The equilibrium constant expression for this reaction would be written as:

$$K = \frac{\left[Na^+(r)\right]\left[H^+(aq)\right]}{\left[Na^+(aq)\right]\left[H^+(r)\right]}$$

For this specific reaction (the exchange of hydrogen ions with sodium ions) the experimental value is K = 0.83. You may also wonder what anions do as they migrate through the resin bed. Can they enter the resin from the aqueous phase? Yes, but only slightly, because the mass of sulfonate groups creates a high negative charge density in the resin that repels most anions. The resin comes in the form of small beads in order to maximize the area in contact with the external solution. The exchange of cations occurs by diffusion within the resin particles. Thus, the smaller the beads, the faster equilibration occurs. Unfortunately, when beads get too small, flow rates get so slow as to be impractical.

In an actual experiment, the ion exchange resin is typically packed into a vertical column (a buret) as shown. A solution is then introduced at the top of the column and collected from the bottom.

Suppose we start with Dowex 50 resin in the hydrogen ion form, where the cations within the resin are all H^+, and equilibrate it with an aqueous solution of sodium chloride. As the solution containing sodium ions (the species in which we are interested) contacts the top of the resin, some of the sodium ions are replaced by an equivalent number of hydrogen ions. The resulting solution moves down the column and encounters a fresh portion of resin where more of the sodium ions are replaced by hydrogen ions. The solution emerging from the bottom contains the same number of moles of hydrogen ions as the number of moles of sodium ions in the initial solution. The original solution is kept moving through the column by the continual addition of distilled water at the top. This technique is called ion exchange chromatography.

We can easily analyze for the hydrogen ion concentration of the final solution by titrating with a standardized base solution using phenolphthalein as the indicator. The hydrogen ion concentration in the final solution is equal to the total concentration of positive charge in the original solution. Thus, through the use of the ion

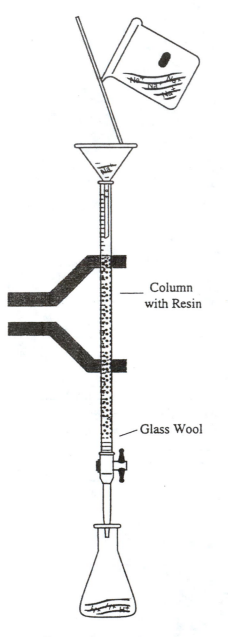

Column with Resin

Glass Wool

Preparing an ion-exchange column.

exchange resin, we have transformed an analysis for cations such as Na^+, which is very difficult, into an analysis for H^+, which is very easy.

LABORATORY PROCEDURES: TOTAL CATION CONTENT

The total cation content of a water sample refers to the total concentration of cations dissolved, predominantly Na^+, K^+, Ca^{+2} and Mg^{+2}, and to a lesser extent Fe^{+2}, Ba^{+2} and Sr^{+2}.

Objective: To determine the total cation concentration by exchanging the H^+ cations of Dowex resin with the cations in the water sample, and subsequently performing an acid-base titration to determine the quantity of H^+ released. For more information about commonly-occurring cations in natural waters, see pages 108 and 114. For more information about the ion exchange method, see page 114.

All glassware used in this experiment should be acid-washed and well rinsed according to the procedures on page 25.

Note

Work in groups of three. Each group member should do the analysis and the group should compare the results obtained. If the three results are not within ±5% of each other, repeat the analysis.

Note

1   Transfer a large test tube full of the ion exchange resin into a 250-mL beaker. Decant any excess liquid. To ensure that the resin has only H^+ attached and no other cations, add approximately 50 mL of 6 M HCl and stir the resin at intervals for about 5 minutes. Decant the liquid into a labeled waste beaker. Add 100 mL of distilled water, stir, and decant. Repeat the distilled water wash until the pH of the rinse water (determined using pH paper) is about 4.

2   Place a small, loose-fitting plug of glass wool in the bottom of a 25-mL buret to prevent resin from blocking the tip. Transfer the resin completely to the buret, using plenty of water from a wash bottle. The resin will be easier to transfer from the beaker to the buret if you do not drain off all the excess water. During this operation, you may open the valve of the buret to drain off excess water, but from this point on keep the level of the liquid at least 1 mL above the top of the resin or air bubbles will form in the buret and you will have to start over. Your resin column should occupy approximately 15 mL of volume.

3   After the resin has been completely transferred and settled in the column, open the stopcock and wash the column with about 30 mL of distilled water at a flow rate of several drops per second. The pH (measured using pH paper) of the last drops should be 4 or greater. Drain the column until the liquid level is just above the surface of the resin bed.

4   Pipet 5 mL of the water sample carefully onto the column. Allow the sample to drain down the inside wall of the column so as not to disturb the surface of the resin bed.

5   Now open the stopcock, and collect the liquid in a 250-mL Erlenmeyer flask. When the liquid level of the sample in the column nears the resin surface, stop the flow and carefully wash the walls of the column with about 1–2 mL of distilled water from the wash bottle, letting it drain down the walls onto the resin bed. Be careful not to disturb

the resin surface any more than necessary. Open the stopcock and let this wash water enter the column. When the liquid level has again drawn near to the resin surface, carefully fill the column with distilled water. Again, be especially careful with the first part of the water you add so that you do not disturb the resin surface. You should collect approximately 50 mL of liquid. Continue collecting sample until the pH of the eluting drop is greater than pH 6 when tested with pH paper. (Why is it not important to record accurately the volume of liquid collected?)

6 Add three to four drops of phenolphthalein indicator to your sample that you collected, and titrate with the standardized NaOH solution to the endpoint where the indicator changes from colorless to a light pink. The three titration volumes obtained by each of the three members of your group should not differ by more than ±5%. If this is not the case, repeat the analysis.

Note

Be sure to write down the EXACT concentration of the standardized NaOH from the bottle you used to fill your buret. You will need to use it in your calculation.

Waste Disposal

The ion exchange resin can be recycled. Discard any neutralized samples and excess base from your buret in the waste container provided. Avoid putting resin in the liquid waste.

Calculations

Calculate the concentration of positive charge in the sample in moles of positive charge per liter of sample. Remember that this number represents the sum of Na^+, K^+, Ca^{+2}, and Mg^{+2}, with trace amounts of Fe^{+2} and Sr^{+2}.

Chapter 8

Atomic Absorption Spectrophotometry

Atomic absorption spectrophotometry (AAS) is an extremely versatile analytical tool. The technique of AAS is used to determine the concentration of trace quantities of metals in a sample. In natural systems, most metals exist as cationic species, either as free cations or as cations that are closely associated with specific anions. Many different metals can be determined using AAS. The periodic chart in Figure 8-1 shows all the elements that can easily be determined by AAS.

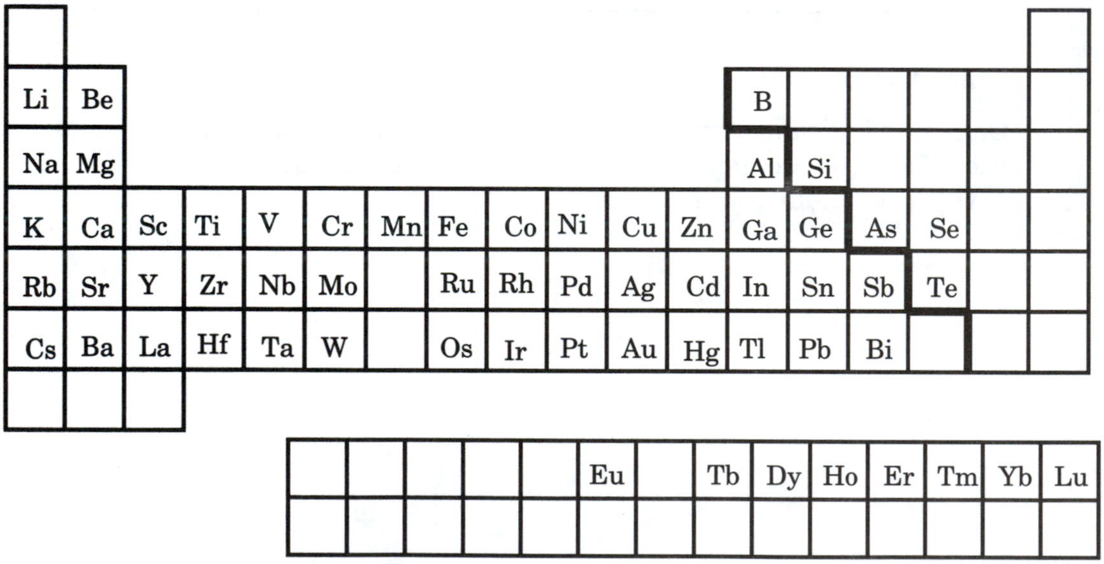

Figure 8-1: Periodic table of elements that can be analyzed by AAS.

HOW DOES AAS WORK?

If you were to place a metal wire in a solution containing sodium ions, then put the wire into a Bunsen burner flame, you would observe a yellow color in the flame. If you were to carry out the same experiment using a solution of lithium ions, you would observe a brilliant red color in the flame. These colors are the result of atomic species (sodium, lithium) absorbing energy from the flame and re-emitting this energy as light. There are two important facts to note here:

Atomic absorption spectro-photometry provides informa-tion about the identity and concentration of cations in so-lution.

1   Each element absorbs and emits a different color (i.e., energy) of light.

2   The more concentrated the solution, the brighter the light emitted.

Atomic absorption spectrophotometry makes use of these properties of elements to measure how much of a particular ion is in a sample.

Some Theory on the Absorption and Emission of Light

Consider the atom, made up of protons and neutrons in the nucleus and electrons in orbitals around the nucleus. The **ground state** of the atom is the state in which all the electrons are in the orbitals which result in the lowest possible energy state. When an atom is provided with an external source of energy, an electron can undergo **excitation** to a higher energy orbital, and the atom is raised to a higher energy state (see Figure 8-2).

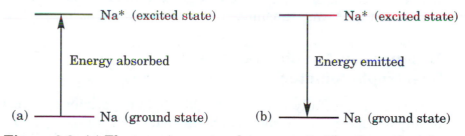

Figure 8-2: (a) Electrons in an atom become excited by absorption of light. (b) Electrons in an atom lose energy by emission of light.

The separation between energy levels is specific to each particular element, with different separations resulting in the absorption or emission of different colors of light for different elements. This energy gap (E) is directly proportional to the **frequency** (ν) of light absorbed or emitted by the element and inversely proportional to the **wavelength** (λ) of light, as related by the speed of light (c) and Planck's constant, h.

When gaseous atoms absorb energy, they are excited to a higher energy state. When the atom returns to the ground state, this energy is re-emitted as light.

$$E = h\nu = \frac{hc}{\lambda}$$

If a beam of light contains photons of an energy that matches the spacing between the two energy levels, some of the incident light will be absorbed by the sample. This absorbed energy causes an electron to be promoted from the ground state to an excited state. The result is that not all of the light that was directed at the sample will be transmitted through to the other side (see Figure 8-3).

Figure 8-3: Only a fraction of the incident light is actually transmitted through the sample.

Transmittance is the ratio of the intensity of transmitted light to incident light.

For a simple analogy, think about how a flashlight beam might behave if you were to shine it through a glass of apple juice. On the other side of the glass, the light would be dimmer than the flashlight beam by itself. The **transmittance**, **T**, of the solution is just the ratio of transmitted light to incident light:

$$T = \frac{I_t}{I_o}$$

where I_o is the intensity of the incident light and I_t is the intensity of the light that was transmitted through the sample.

Design of an Atomic Absorption Spectrophotometer

If you can successfully visualize the flashlight experiment above, it is only necessary to add some sophisticated electronic components to design an AAS. A schematic of the instrument is shown in Figure 8-4 below.

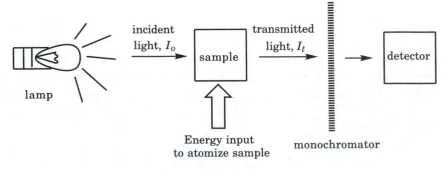

Figure 8-4: Schematic of an atomic absorption spectrophotometer.

A lamp that gives off light of the appropriate energy (wavelength) is passed through a sample that contains the atomic species in the **gas phase**. The sample absorbs some of the incident light, transmitting the remainder. The light then passes through a **monochromator**, a device that further "purifies" the light so that only light with an extremely small range of energies (or wavelengths) passes through. Think of a monochromator as a filter that excludes all but the desired wavelength of light. The light is then passed into a **detector**, a sensitive electronic device that can accurately measure the intensity of a beam of light. The amount of light absorbed is directly proportional to the number of absorbing atoms in the path of the light.

To increase precision and reduce interferences, element-specific lamps are used to produce light of a very specific wavelength.

Flame Atomization

In order to get the atoms in a sample into the gas phase, it is often necessary to transform any ionic forms of that element from *charged* species in solution to *neutral,* gaseous atomic species. This typically requires the input of a large quantity of energy in the form of heat. For **flame AAS**, the heat source is a very hot air/acetylene flame.

The processes that occur in the flame for sodium ions (Na^+) in an aqueous solution of sodium chloride (NaCl) are outlined in Figure 8-5. Step 6b is where energy from a lamp shining on the sample is absorbed by the gaseous atoms. Some atoms are also excited by absorbing heat energy from the flame (step 6a); this process can interfere with the absorption of light from the lamp and a variety of methods have been developed to reduce the contribution of the flame to the total absorption of energy. **Matrix modifiers** are often used to prevent the loss of instrumental sensitivity resulting from excitation of atoms by the heat of the flame.

Matrix modifiers enhance the sensitivity of an analysis.

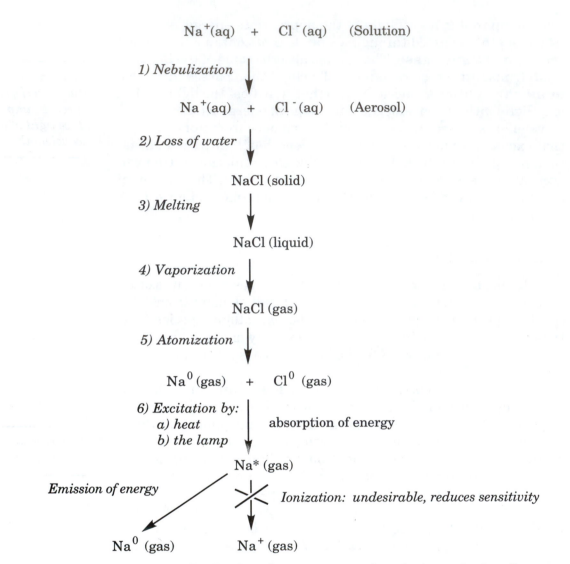

Figure 8-5: Production of gaseous atoms from ionic species in a flame proceeds through several steps.

Relating the Amount of Light Absorbed to Concentration: Beer's Law

Before AAS can be used to determine the concentration of a species in solution, the instrument must be **standardized** to determine the relationship between the amount of light absorbed by a given quantity of the substance being measured. Modern instruments usually measure the **absorbance (A)** of a sample, not transmittance, where the relationship between absorbance and transmittance is a logarithmic one.

$$A = -\log T$$

Absorbance is linearly related to concentration by **Beer's Law:**

$$A = \mathcal{E}bc$$

where A is the absorbance (a unitless number), b is the path length of the light through the sample (typically ~10 cm for flame AAS analysis), c is the concentration of the substance in solution, and ε (the Greek letter *epsilon*) is a constant that is specific for a particular substance and is determined experimentally for each species analyzed. The higher the concentration of the substance in solution, the greater the absorbance that is measured by the instrument. This relationship is typically linear only over a limited range of concentrations, deviating from linearity at higher concentrations because of interactions between absorbing species. A typical Beer's Law plot, also known as a **standard curve**, is shown in Figure 8-6.

The amount of light absorbed by a substance is proportional to its concentration.

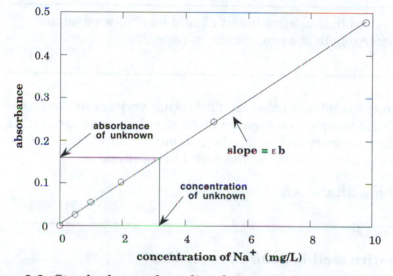

Figure 8-6: Standard curve for sodium by atomic absorption spectrophotometry.

To construct a standard curve, it is necessary to make up solutions of known concentrations, measure their absorbances, and plot the results as in Figure 8-6. The slope of the line is εb. The concentration of the ion of interest in an unknown solution can then be determined by using this value of the slope and Beer's law to calculate a concentration from an absorbance reading. The actual value of εb will depend on experimental conditions such as the flame temperature and the concentration of matrix modifier, and as a result, the instrument must be restandardized for each analysis.

LABORATORY PROCEDURES: CATIONS BY ATOMIC ABSORPTION SPECTROMETRY

Objective: To determine the concentrations of sodium (Na^+), potassium (K^+), calcium (Ca^{+2}), and magnesium (Mg^{+2}) in the sample using flame atomic absorption spectrophotometry. For detailed information on the theory of atomic absorption spectrophotometry, see pages 120–125. More information about sources of these cations in natural waters can be found on pages 109 and 114.

A *filtered* sample should be used for atomic absorption analysis.

All glassware used in this experiment should be acid-washed and well rinsed according to the procedures on page 25.

The atomic absorption spectrophotometer is capable of detecting very small quantities of metals. It works by atomizing the sample in a high temperature flame and measuring the absorbance of light of a specific frequency characteristic to a particular element, where the absorbance is proportional to the amount of dissolved metal in solution.

How to Organize Your Time with the AA

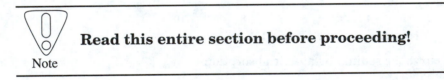

Read this entire section before proceeding!

You will be working in teams for this analysis. When you get to the AA, there are four tasks that need to be done:

- Calibrate the instrument.
- Plot out the standard curve to check that the standards are OK.
- Do a quick test of your sample to determine if you must dilute it.
- Mix your samples with matrix modifier and run them.

Calibrate the Instrument

Using the instructions for the instrument provided by your instructor, set up the instrument parameters and run the standards, making sure to write down the absorbance reading for each standard. Make a table in your lab notebook in which you record concentration and absorbance values for each standard. These will be your x (concentration) and y (absorbance) values for the plot of your standard curve.

Plot the Standard Curve

Using the values you measured for absorbance as a function of concentration, create a plot of absorbance vs. concentration. Does your data make a straight line? If not, recalibrate the instrument.

Do a Quick Test of the Sample to Determine If You Must Dilute It

Before adding any matrix modifier to the sample, pour a few milliliters into a clean vial, insert the nebulizer tube into the vial and measure the absorbance of the sample. If the absorbance reading is higher than that of the highest standard, the sample must be diluted to an appropriate level, using pipets and a volumetric flask (see instructions in Appendix B). Be sure to note the dilution factor in your notebook and take it into account when you evaluate your data.

> **Note**
>
> Even though you receive a reading that is off scale, the instrument will provide an approximate concentration—use this number to guess the best dilution to make to bring the concentration of the sample into the midrange of the standards.

Mix the Sample with Matrix Modifier and Run It

After diluting the sample appropriately, follow the procedure below to mix the samples with the appropriate matrix modifier. Work as a team to share the responsibilities of preparing the samples. Mix the samples well, then run them and record the absorbance reading from the AA.

1 **For sodium analysis:** Cesium chloride must be added in order to maximize absorption by this ion. For each sample, pipet 8 mL of sample and 2 mL of CsCl solution into a vial. Cap the vial and shake well to mix.

2 **For potassium analysis:** Cesium chloride must be added in order to maximize absorption by this ion. For each sample, pipet 8 mL of sample and 2 mL of CsCl solution into a vial. Cap the vial and shake well to mix.

3 **For calcium analysis:** Potassium chloride must be added in order to maximize absorption by this ion. For each sample, pipet 8 mL of sample and 2 mL of KCl solution into a vial. Cap the vial and shake well to mix.

4 **For magnesium analysis:** Lanthanum chloride must be added in order to maximize absorption by this ion. For each sample, pipet 8 mL of sample and 2 mL of $LaCl_3$ solution into a vial. Cap the vial and shake well to mix.

> **Waste Disposal**
>
> The solutions from these analyses may be drain disposed in most locations. Check with your instructor.

Calculations

1   Determine the concentration of each ion measured in moles per liter. Be sure to take into account any dilutions you made.

2   Determine the total moles of positive charge in the solution by taking the concentration of each ion and multiplying it times the charge on the ion. Sum these numbers to obtain the total moles of positive charge in the sample.

Chapter 9

Data Analysis

Measuring the concentrations of environmental pollutants is a tricky business. There are many ways to make errors in the measurements, including errors on the part of the analyst in weighing and measuring as well as errors in the calibration and operation of the instruments. In addition, there will always be some random error in any sampling or analytical process. As a result, it is *essential* to include an estimate of the uncertainty associated with each analysis. Remember that *all* measurements have some degree of uncertainty. Estimating uncertainty is not always a straightforward process; however *data that are reported with no uncertainty are worthless!* In contrast, results that have a large uncertainty associated with them may still be quite useful if the uncertainty of the measurement is known.

PRECISION AND ACCURACY

There are two types of uncertainty that must be considered, precision and accuracy. **Precision** reflects the reproducibility of measurements made by identical methods. For example, to find the precision of anion analysis by ion chromatography, you would carry out the IC analysis on two (or more) portions of the same sample and check to see how similar the two results are to each other.

Accuracy is a measure of how close a measurement is to the "actual" value. An atomic clock measures time more accurately than a quartz clock, which is in turn more accurate than a spring-wound kitchen timer. In environmental settings, we often cannot know the "actual" value. The best we can do then is to be sure our method is giving reliable answers by checking the method on a substance that is prepared in such a manner that we *know* what value the instrument should provide. Such a substance is called a **standard**. Standards can help assess both precision and

Precision provides information about the reproducibility of the measurement, whereas accuracy is representative of how close the measurement is to the true value.

accuracy and are essential in verifying the calibration of instruments.

Precision and accuracy do not necessarily go hand in hand. Measurements that are very precise are not necessarily very accurate, and vice-versa. For example, a digital pH meter may be very precise in reporting that the pH of a pond is 6.06, rather than 6.05 or 6.07, but how accurate is that measurement? Instruments must be calibrated against known standards to insure accuracy. If that pH meter had not been recently calibrated, or the temperature of the pond was different from the temperature at which the meter was calibrated, or the meter had not been allowed to warm up, or a dozen other details had not been considered, then the actual pH of the pond might be significantly different—like 6.12 or even 4.96!

The use of a universal pH indicator solution on a sample from that same pond might give a greenish color, indicating that the pH is 7. The *accuracy* of the indicator solution is not in question; the color indicates that the pH was not 5, 6, or 8, but rather 7. However, the *precision* of the measurement made with the indicator solution is not very good, because the uncertainty associated with most indicators is quite large (between 0.5 and 1.0 pH units). We do not know if the pH is 7, 7.1, 6.8, or even 6.93.

The game of darts provides another example of the difference between precision and accuracy (Figure 9-1). If the objective of the game is to hit the bull's-eye in the center of the target, then both accuracy and precision are required in throwing the darts.

A measurement that is very precise is not necessarily accurate.

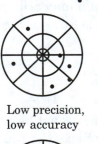
Low precision,
low accuracy

Low precision,
high accuracy

High precision,
low accuracy

High precision,
high accuracy

Figure 9-1: Precision and accuracy illustrated by dart throwing.

If a person throws all of their darts in the outer border of the number 1 wedge, then they are very precise (and perhaps very good at a different game of darts), but not very accurate. Although all of their darts have ended in the same place, that place is far from their goal. In the same game of darts, if the thrower is accurate, but not precise, their "average" throw will be in the center of the target, but it is possible that no single throw will hit the bull's-eye. Just as we derive little benefit from numbers that are neither precise nor accurate, the imprecise, inaccurate darts thrower has little chance of victory.

Sources of Error

The sources of error that lead to inaccuracy in a given measurement are categorized as either **statistical errors** or **systematic errors**. Statistical errors, also called **random errors**, have an equal probability of making the reported measurement too high or too low. Statistical errors may be minimized by repeating a measurement several times. Small statistical errors are often associated with high precision. Systematic errors, or **determinate errors**, occur in the same direction each time. If one weighs a series of soil samples and forgets to zero the balance at the start of the measurements, then a systematic error has been introduced. Impurities in water used in the dilution of aqueous environmental samples might lead to large systematic errors in determining ion concentrations. Systematic errors are much more difficult to quantify than statistical errors. Each time you perform a measurement, you should consider all the possible sources of error and work to minimize those that will significantly reduce the precision or accuracy of your measurement.

Systematic errors usually indicate a problem with equipment or technique.

Significant Figures

The precision or uncertainty in a measurement can be expressed by the number of significant figures used to express a result. More precise results have more significant figures, indicating there is little spread in the data. In most cases, the number of significant figures a number has is merely the number of digits that number has. Confusion may arise when zeroes are included. Leading zeroes are not significant, and trailing zeroes are only significant if the number contains a decimal point. Table 9-1 helps to illustrate how to determine the number of significant figures.

"Exact" numbers are considered to have an infinite number of significant figures. The number π (3.141592653589793...) and the number e (2.7182818...) are examples of noninteger, exact numbers. Integers such as 10 fingers, 13 cookies, or 32 students, are also considered exact, because they are obtained by counting, rather than by measuring.

When it is necessary to perform a calculation, the number of significant digits in the reported result depends on the operation performed. Some simple rules are helpful guides: For addition and subtraction the answer has the same number of *decimal places* as the *least* precise measurement. For multiplication or division, the

Table 9-1: How Many Significant Figures?

Number	Significant Figures	Number	Significant Figures
135	3	72	2
13500	3	720	2
13500.	5	720.	3
1.350×10^4	4	7.2×10^2	2
0.00135	3	0.072	2
0.0013500	5	0.0720	3
1.35×10^{-3}	3	7.2×10^{-2}	2
1.3500×10^{-3}	5	7.20×10^{-2}	3

For maximum accuracy, round the number only after the last calculation is performed.

answer has the same number of *significant figures* as the *least* precise measurement. When a series of calculations are performed, do not round off until the final result is obtained. If you are rounding a number that ends in a five, round it to the closest *even* digit. For example, in a case where the thens digit is the most significant, 11.35 would round to 11.4 and 1.25 would round to 1.2. Table 9-2 contains a few illustrative examples. Note that the use of scientific notation leaves no doubt as to how many significant figures a number contains. Reporting the result of 3120 - 17 would be confusing if it were not expressed in scientific notation.

Table 9-2: Significant Figures from Mathematical Operations

Calculation	Result	Reported Result	Comment
17 + 3120	3137	3140 or 3.14×10^3	The tens digit is least significant
3120 - 17	3103	3.10×10^3	
3120 × 17	53040	53000 or 5.3×10^4	17 has only two significant figures
3120/17	183.529...	180 or 1.8×10^2	

Significant figures for numbers reported as logarithms are the number of digits after the decimal place.

The number of significant figures in a number that is the logarithm of another number is the *number of digits after the decimal place*. The number of digits before the decimal place give the order-of-magnitude of the number, not its precision, and thus are not counted as significant figures. This is particularly important in pH measurements. Remember that pH = $-\log_{10}[H_3O^+]$. Table 9-3 contains some numerical examples of the reporting of significant figures in pH measurements.

Table 9-3: Significant Figures Used in Reporting pH

[H_3O^+]	pH	Significant Figures
1×10^{-7}	7.0	1
1.0×10^{-7}	7.00	2
1.1×10^{-7}	6.96	2
1.2×10^{-7}	6.92	2
1.1×10^{-11}	10.96	2
1.15×10^{-11}	10.939	3

Examination of some of the examples in Table 9-3 shows the rationale for the rules in reporting pH. If we reported that a solution with [H_3O^+] = 1.1×10^{-7} M was pH 7.0, rather than pH 6.96, we would not be able to distinguish it from a solution with [H_3O^+] = 1.0×10^{-7}, and we would reduce the precision of our measurement.

The most important point when considering significant figures for the reporting of data measured in the laboratory is that *the number of significant figures should reflect the precision of the measurement*. A good rule is that, unless otherwise stated, if a number is reported without uncertainty limits, then there is an uncertainty of ±1 in the least significant digit (the rightmost digit). Some feel that an uncertainty of ±2 in the least significant digit is more appropriate.

The number of significant figures should reflect the precision of the measurement.

Some Simple Statistics

One way to assess the uncertainty of your measurements is to carry out a statistical analysis on the data. Statistics allow us to measure properties of a *body* of data, providing a measure of its reliability. The computer program you will use to process the data will carry out a variety of statistical analyses for you; however, you should go through at least one calculation of each type so you see for yourself how to do it. First, some definitions.

Average (also called **mean**): The number calculated by adding all the values of a set of data together and dividing by the number of measurements (*N*). Frequently abbreviated by the symbol shown below (pronounced "*x*-bar").

$$\overline{x}$$

Median: The middle result of the data when it is arranged by increasing or decreasing numerical value. When a data set contains a few values that are extremely different from the others, the average is often not very representative of the true value. In this case, the median provides a more accurate representation of the data set.

Standard Deviation: Representative of the average amount the readings deviate from the average of the data set. This number provides a measure of the spread of the data around an average value. The standard deviation (abbreviated s) can be calculated using the following equation:

$$s = \sqrt{\frac{\sum_{i=1}^{N}(x_i - \bar{x})^2}{N-1}}$$

where N is the number of measurements in the data set, x_i is the individual measurement for which you are calculating the standard deviation, and $\bar{x}$ is the average value obtained for the data set.

Relative Standard Deviation: Frequently, standard deviations are reported as a percentage of the average value. This value, called the relative standard deviation (abbreviated RSD), can be calculated using the following equation:

$$\text{RSD} = \frac{s}{\bar{x}} \times 100\%$$

Example Calculations

Given the following set of data for the sulfate concentrations in water samples, we will calculate the average (mean), standard deviation, and relative standard deviation for the data.

Sample No.	Sulfate Concentration (ppm)
1	2.035
2	2.192
3	1.998
4	2.005
5	2.067
6	2.154

For this data set, the number of samples $N = 6$. The **average** value of the sulfate concentration for these samples is

$$\bar{x} = \frac{(2.035 + 2.192 + 1.998 + 2.005 + 2.067 + 2.154)}{6} = 2.075$$

For the **standard deviation**, we first need to know the sum of the squares of the deviation of each number from the mean, In the case of these data, this term is equal to

$$\sum_1^6 (x_i - \bar{x})$$

$$(2.035 - 2.075)^2 + (2.192 - 2.075)^2 + (1.998 - 2.075)^2 + (2.005 - 2.075)^2$$
$$+ (2.067 - 2.075)^2 + (2.154 - 2.075)^2$$
$$= (-0.04)^2 + (0.117)^2 + (-0.077)^2 + (-0.07)^2 + (-0.008)^2 + (0.079)^2$$
$$= 0.0016 + 0.0137 + 0.0059 + 0.0049 + 0.0001 + 0.0062$$
$$= 0.0324$$

Next, this result is divided by 5 ($= N - 1$) and the square root is taken

$$s = \sqrt{\frac{0.0324}{5}} = \sqrt{0.00648} = 0.081$$

This standard deviation of 0.081 indicates that this data is uncertain in the hundreths place and should be reported as ± 0.08 For example, the first data point is 2.04 ± 0.08). This gives an idea of the reliability of the numbers.

The **relative standard deviation** for this data set would be

$$RSD = (s / \bar{x}) \times 100\%$$
$$= \frac{0.081}{2.075} \times 100\%$$
$$RSD = 3.9\%$$

That is, the reliability of these numbers is ±3.9%.

Should You Throw Out That Data Point?

There will certainly be instances where one measurement looks very out of place within a given set of data. For example, perhaps the amount of chloride found in the water supply is typically 5.6 ppm. If only one of the samples (out of 20) gave a reading of 25.2 ppm for chloride, it would be worth worrying about whether to include that data point in the data set used to calculate the average and the standard deviation.

If one were to calculate the deviation of each measurement from the mean and plot how often each value occurred versus the

$$(x_i - \bar{x})$$

deviation, you would obtain a plot similar to that shown below in Figure 9-2, called a **histogram**. You might be familiar with this as

Discarding data points should not be done without careful consideration and statistical analysis of the error.

the "curve" that a professor uses to scale the grades in a class. This curve represents a **Gaussian distribution** (the name of the mathematical function that describes the curve) of the deviations of the measurements from the mean. The graph shows that the maximum number of data points occurs at the mean value of the measurement, with fewer data points observed at values of the measurement distant from the mean. For a Gaussian distribution of deviations, it is known that 68% of the measurements will fall within one standard deviation ($\pm 1s$) of the mean and 95% will fall within two standard deviations ($\pm 2s$) of the mean. This theory is only truly accurate for an infinite number of data points where the only source of error is random.

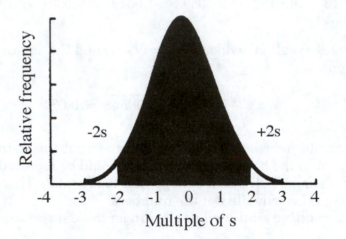

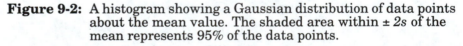

Figure 9-2: A histogram showing a Gaussian distribution of data points about the mean value. The shaded area within $\pm 2s$ of the mean represents 95% of the data points.

One method for deciding which data to keep and which to throw away is to do the following:

1 Calculate the average of the measurements *including* the questionable data point.

2 Calculate the standard deviation for the data set. (Computer programs such as Excel or even hand calculators will do this for you in a matter of seconds!)

3 Assess whether the questionable data point is within *two standard deviations* of the mean.

4 If it is, you should keep the data point. If not, throw it out and recalculate a new average and standard deviation based on the remaining data. Be sure to mention in your report what you did.

On the following page is an exercise to be done during class to help cement some of the principles of error analysis in your mind.

LABORATORY PROCEDURES: A STATISTICS EXERCISE

Your Mission

When you read a number off an instrument display or when you obtain a number of milliliters for a titration volume that is representative of the concentration of a substance in your water sample, how many significant figures should you keep? Is the instrument or your titration truly accurate enough to provide concentrations to the third decimal place?

On the following page in Figure 9-3, you are provided with the results of six sequential injections of a multicomponent standard into an ion chromatograph, an instrument that measures the concentration of ions in aqueous solution. The actual concentrations (assuming the standard is good) of the ions are given in the top row. The purpose of this exercise is to determine the uncertainty in measurements, i.e., if the instrument reports a value of sulfate as 4.567 mg/L, is it *really* 4.567 mg/L of sulfate or is it better reported as 4.6 mg/L? This exercise will help you determine how many significant figures to keep in reporting data.

How to Proceed

Read the section on *Data Analysis* on pages 129–136. In your laboratory notebook, do the following for each ion (F^-, Cl^-, NO_3^-, PO_4^{-3}, and SO_4^{-3}):

1 Calculate the average value from the six sample runs.

2 Calculate the standard deviation, *s*.

3 Calculate the uncertainty in the measurement, defined as *2s*.

4 Calculate the relative standard deviation (RSD).

Your instructor may encourage you to work in groups, but *each person* should go through *every one* of the steps at least once to see how the calculations are done. A good way to do this is for each person to do the calculations for one ion.

Using the Results

Answer the following questions in your laboratory notebook.

1 For this instrument, how should you report a value of 4.325 mg/L of phosphate, i.e., how many significant figures should you keep to reflect the true accuracy of the measurement?

2 Is there random error in the data? Systematic error? Explain.

3 Which of the reported ion concentrations is more precise? Explain your reasoning.

F^-	or	Cl^-
NO_3^-	or	PO_4^{-3}
PO_4^{-3}	or	SO_4^{-2}

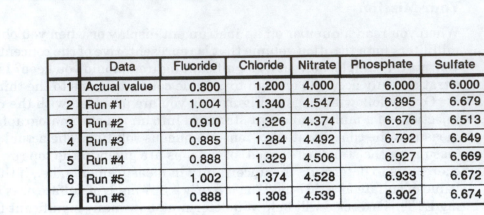

File: MANAN171.D01 Sample: KTEST

	Data	Fluoride	Chloride	Nitrate	Phosphate	Sulfate
1	Actual value	0.800	1.200	4.000	6.000	6.000
2	Run #1	1.004	1.340	4.547	6.895	6.679
3	Run #2	0.910	1.326	4.374	6.676	6.513
4	Run #3	0.885	1.284	4.492	6.792	6.649
5	Run #4	0.888	1.329	4.506	6.927	6.669
6	Run #5	1.002	1.374	4.528	6.933	6.672
7	Run #6	0.888	1.308	4.539	6.902	6.674

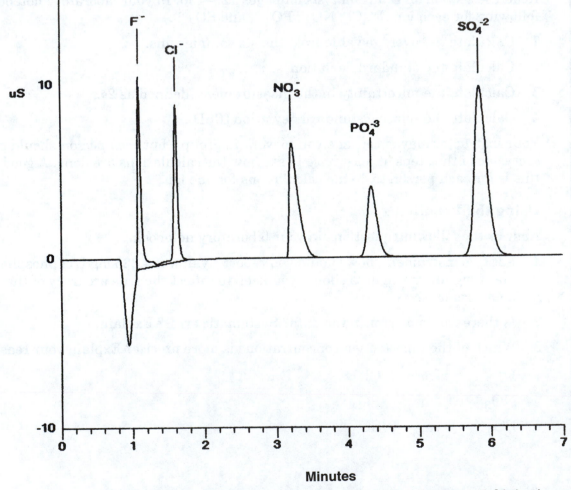

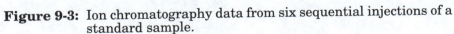

Figure 9-3: Ion chromatography data from six sequential injections of a standard sample.

LABORATORY PROCEDURES: DATA ANALYSIS WORKSHEET

Names of Group Members:

Spend 15–20 minutes to take a close look at the data your class collected. Evaluate the class data (and the data you, personally, collected) in the context of the following:

- Is there agreement between replicate samples?

- Are there any data points that are very different than what you'd expect based on the rest of the data? If so, are they off by a particular factor such as a dilution factor?

- Check the calculations made by each group member. Are they correct?

- Do the data make sense in terms of the location of the sample, i.e., if a sample has a high concentration of a substance, is there a reason to expect to find a high concentration at that location?

  Group 1: Total and fecal coliform
  Group 2: Dissolved oxygen and conductivity (field or laboratory measurements)
  Group 3: pH and alkalinity
  Group 4: Total cation content, calcium, magnesium, sodium, potassium
  Group 5: Fluoride, chloride, nitrate, phosphate, sulfate

Working with your group, answer the following questions in your laboratory notebook.

1 List all missing data points. If some of the missing data are from someone in your group, provide the missing data here.

2 List all suspicious data points and speculate as to a possible reason for their lack of agreement with what you would expect. Do not forget to look at the map to draw correlations with the site.

Turn in your notes to the instructor after class discussion of the data.

FINAL REPORT FOR THE WATER MODULE

The report should include the following parts:

I.   Introduction
II.   Experimental Procedures
III.   Calculations
IV.   Results and Discussion
V.   Conclusions
VI.   Appendices (see details following)
VII.   References
VIII. Questions

The report should be about 5–10 pages long, excluding any graphs or appendices.

I. Introduction

The introduction should provide some background information on the study site and a discussion of *why* the study was carried out. Think about what a city official might like to know in order to make a decision about whether or not the water source you have analyzed should be used for drinking water. Include enough information about the study site so that a reader who is unfamiliar with the area will know what you are talking about. You should also include some information on what makes water drinkable or not and what you might expect to find as "natural" constituents of fresh water.

II. Experimental Procedures

The experimental section of your report should contain information about how samples were collected, including the container type and how it was cleaned. If the recent weather might have affected your results, it should be included as well. The experimental section is the place in your report where it is most appropriate to include a map, with the sample sites *clearly* labeled on the map. For the laboratory procedures, provide a *brief* description of what was done, not a detailed, blow-by-blow explanation of every excruciating detail. The whole procedure section should be *no longer than* (and possibly shorter than) two pages, depending on the analyses you carried out.

III. Calculations

- Include any calculations you did for the analyses you carried out.

- Compare the values obtained for the total moles of cations (Na^+, K^+, Ca^{+2}, and Mg^{+2}) per liter and the total moles of anions (HCO_3^-, CO_3^{-2}, OH^-, Cl^-, F^-, NO_3^-, PO_4^{-3}, and SO_4^{-2}) per liter. Do they match? If not, why not?

IV. Results and Discussion

In this section, the data from the entire class should be presented and the implications of the findings discussed. This is one part of the lab where your creativity will shape what you decide to do with the data. The very least you need to do with the data for each analysis is report the results as a function of location, discuss your confidence in the results, compare the value with typical values for natural waters and discuss the implications of using the water source as drinking water. Use the map of the site to locate the sample site and be sure you have commented on the possible relationship of the site to the concentration of the different components of the water sample. Additionally, try to think about possible relationships that might exist in the data, such as the relationship of the concentrations of different ions to each other. Correlation plots of the concentration of one ion vs. that of another can show relationships between ions that provide information about the source of the ions in the water supply.

The data set you will be analyzing might be quite large and complex, depending on how many analyses you have carried out on the samples. Here are some suggestions for making the data set understandable and useful.

- You may wish to consolidate the data into groups of similar types, e.g., consider averaging samples from similar sites to increase the statistical significance of your numbers. If you choose to do this, be *very clear* about how and why you grouped samples as you did. Make a new table to clearly show the groupings and do not forget to report the statistics on the groups of numbers. Excel can help you do this easily.

MOST IMPORTANT!!! Any number obtained by averaging the data **must be accompanied by the standard deviation.** Otherwise, it is meaningless.

Note

- Use the map of the site as a means to transmit the data and be sure to point out the relationship of the site to the concentrations of the different species in the water sample.

- Use graphs to depict your results and to establish relationships between parameters that were measured. **Column** graphs are useful to show the distribution of a substance at different sites. **Scatter** plots are useful to determine if there is a connection between two different measured quantities, for example, $[Na^+]$ and $[Cl^-]$ or $[Ca^{+2}]$ and $[CO_3^{-2}]$. Pie charts or any kind of connect-the-dots plots are not particularly useful and should not be used.

If you make new tables of data or if you include graphs, *it is essential to discuss them*. Do not include a graph or a new table and not say anything about it. If you eliminate data points, be very clear about the reasons you chose to discard the data. Be sure to LABEL all graphs and tables, giving them a number and an informative title. Label the axes so it is clear what information you are trying to transmit with the graph. Attach all graphs to the end of the report. In the report, refer to "Table 1" or "Figure 3" so the reader will be able to follow the discussion.

There are a nearly limitless number of graphs you *might* plot, but not all of them will give you any useful information. Choose the best ones by looking at them on the computer screen, but limit the number of graphs included in your final report to those which provide useful information about the site.

Perhaps the most important part of your report will be to compare your data to known values, including expected values for freshwater systems and the Federal Water Quality Standards (see p. 17). Inclusion of these values in a column graph of your data is a particularly informative way of presenting the comparison. Speculation on the source of any contaminants and possible ways to reduce these contaminants is also an important aspect of the report.

IV. Conclusions

The conclusion section of your report is a good place to reiterate the problem and summarize the data that support your conclusion about whether or not the water source should be used as a drinking water supply. The conclusion section should also discuss the measures that would be necessary to bring the water supply into compliance with the drinking water standards. It need not be long, just to the point.

VI. Appendices

Include the appendices chosen by your instructor:
a   A discussion in your own words of pH and alkalinity
b   A discussion in your own words of how the dissolved oxygen meter works
c   A discussion in your own words of the chemistry behind the Winkler titration
d   A discussion in your own words of how a colorimetric method works
e   A discussion in your own words of how ion chromatography works
f   A discussion in your own words of how atomic absorption spectrophotometry works

VII. References

It is very important to acknowledge all sources from which information was obtained in writing any scientific paper. The lab manual can serve as your primary source material, but you might also wish to seek out and use other books or texts. A proper reference for a book should include the name of the author(s), the title, the publisher and year and place of publication. For a journal article, the reference should include the name of the author(s), the *journal* title and volume number, the year in which this issue was published, and the number of the page on which the article begins.

VIII. Questions

Include the answers to any questions posed in the analyses you carried out.

Lab Reportese: A Note on Grammar and Style

The scientific style of writing attempts to convey what was done to the reader using the least number of words with the greatest clarity. Although it is perhaps not as exciting as the style you might use for an English paper, it is very useful in writing scientific reports and certainly contributes to whether or not the reader will take you seriously.

In general, use past tense, passive voice, third person to describe what was done. Remember, you are describing *what you did*, not giving instructions to someone else. For example:

Correct: "The samples were collected at the Safeway in Oakland."
Incorrect: "While we were out cruising, we collected the samples at Safeway."

Correct: "The sample was dropped and as a result, no data were obtained."
Incorrect: "I dropped the sample all over the floor and couldn't do any of the tests."

Correct: "The GC was calibrated using a set of four standards."
Incorrect: "Calibrate the GC by using a set of four standards."

The incidence of I's, we's, you's, our's, your's, and my's should be very limited in a lab report and may only be appropriate in your final recommendation of whether or not pesticides should be regulated differently. For example, "In light of the data collected, I can not recommend this water supply as safe for drinking without extensive treatment."

Avoid sensationalism. Back up any judgements with data and scientific facts and assess carefully whether you actually have enough data to come to a conclusion. The conclusion "Further study is needed" is also valid.

Appendix A

Ionic Substances in Natural Waters

Table A-1: Principal Chemical Constituents in Water: Their Sources, Concentrations, and Effects Upon Usability

Constituent	Major Sources	Concentration in Natural Water	Effect Upon Usability of Water
Carbonate (CO_3^{-2})	Limestone, dolomite	Commonly 0 mg/L in surface water; commonly less than 10 mg/L in ground water. Water high in sodium may contain as much as 50 mg/L of carbonate.	Upon heating, bicarbonate is changed into steam, carbon dioxide, and carbonate. The carbonate combines with alkaline earths—principally calcium and magnesium—to form a crustlike scale of calcium carbonate that retards flow of heat through pipe walls and restricts flow of fluids in pipes. Water containing large amounts of bicarbonate and alkalinity are undesirable in many industries.
Bicarbonate (HCO_3^-)		Commonly less than 500 mg/L; may exceed 1000 mg/L in water highly charged with carbon dioxide.	
Sulfate (SO_4^{-2})	Oxidation of sulfide ores; gypsum; anhydrite; industrial wastes	Commonly less than 1000 mg/L except in streams and wells influenced by acid mine drainage—as much as 200,000 mg/L in some brines	Sulfate combines with calcium to form an adherent, heat-retarding scale. More than 250 mg/L is objectionable in water in some industries. Water containing about 500 mg/L of sulfate tastes bitter; water containing about 1000 mg/L may be cathartic.
Chloride (Cl^-)	Chief source is sedimentary rock (evaporites); minor sources are igneous rocks. Ocean tides force salty water upstream in tidal estuaries	Commonly less than 10 mg/L in humid regions; tidal streams contain increasing amounts of chloride (as much as 19,000 mg/L) as the bay or ocean is approached. About 19,300 mg/L in seawater; and as much as 200,000 mg/L in brines.	Chloride in excess of 100 mg/L imparts a salty taste. Concentration greatly in excess of 100 mg/L may cause physiological damage. Food processing industries usually require less than 250 mg/L. Some industries—textile processing, paper manufacturing, and synthetic rubber manufacturing—desire less than 100 mg/L.
Fluoride (F^-)	Amphiboles (hornblende), apatite, fluorite, mica	Concentrations generally do not exceed 10 mg/L in ground water or 1.0 mg/L in surface water. Concentrations may be as much as 1,600 mg/L in brines.	Fluoride concentration between 0.6 and 1.7 mg/L in drinking water has a beneficial effect on the structure and resistance to decay of children's teeth. Fluoride in excess of 1.5 mg/L in some areas causes "mottled enamel" in children's teeth. Fluoride in excess of 6.0 mg/L causes pronounced mottling and disfiguration of teeth.
Nitrate (NO_3^-), reported as N	Atmosphere; legumes, plant debris, animal excrement, nitrogenous fertilizer in soil and sewage	In surface water not subjected to pollution, concentration of nitrate may be as much as 5.0 mg/L but is commonly less than 1.0 mg/L. In groundwater the concentration of nitrate may be as much as 1,000 mg/L.	Water containing large amounts of nitrate (more than 100 mg/L) is bitter tasting and may cause physiological distress. Water from shallow wells containing more than 45 mg/L has been reported to cause methemoglobinemia in infants. Small amounts of nitrate help reduce cracking of high-pressure boiler steel.

Constituent	Major Sources	Concentration in Natural Water	Effect Upon Usability of Water
Dissolved Solids	The mineral constituents dissolved in water constitute the dissolved solids.	Surface water commonly contains less than 3000 mg/L; streams draining salt beds in arid regions may contain in excess of 15,000 mg/L. Ground water commonly contains less than 5000; some brines contain as much as 300,000 mg/L.	More than 500 mg/L is undesirable for drinking and many industrial uses. Less than 300 mg/L is desirable for dyeing of textiles and the manufacture of plastics, pulp paper, rayon. Dissolved solids cause foaming in steam boilers; the maximum permissible content decreases with increases in operating pressure.
Silica (SiO_2)	Feldspars, ferromagnesium and clay minerals, amorphous silica, chert, opal	Ranges generally from 1.0 to 30 mg/L, although as much as 100 mg/L is fairly common; as much as 4,000 mg/L is found in brines.	In the presence of calcium and magnesium, silica forms a scale in boilers and on steam turbines that retards heat; the scale is difficult to remove. Silica may be added to soft water to inhibit corrosion of iron pipes.
Iron (Fe)	Natural sources: Igneous rocks such as Amphiboles, ferromagnesian, micas, ferrous sulfide (FeS), ferric sulfide or iron pyrite (FeS_2), magnetite (Fe_3O_4); sandstone rocks such as oxides, carbonates, and sulfides or iron clay minerals. Manmade sources: Well casing, piping, pump parts, storage tanks, and other objects of cast iron and steel that may be in contact with the water	Generally less than 0.50 mg/L in fully aerated water. Ground water having a pH less than 8.0 may contain 10 mg/L; rarely as much as 50 mg/L may occur. Acid water from thermal springs, mine wastes, and industrial wastes may contain more than 6000 mg/L.	More than 0.1 mg/L precipitates after exposure to air; causes turbidity, stains plumbing fixtures, laundry and cooking utensils, and imparts objectionable tastes and colors to foods and drinks. More than 0.2 mg/L is objectionable for most industrial uses.
Manganese (Mn^{+2})	Manganese in natural water probably comes most often from soils and sediments. Metamorphic and sedimentary rocks and mica biotite and amphibole hornblende minerals contain large amounts of manganese.	Generally 0.20 mg/L or less. Groundwater and acid mine water may contain more than 10 mg/L. Reservoir water that has "turned over" may contain more than 150 mg/L.	More than 0.2 mg/L precipitates on oxidation; causes undesirable tastes, deposits on foods during cooking, stains plumbing fixtures and laundry, and fosters growths in reservoirs, filters, and distribution systems. Most industrial users object to water containing more than 0.2 mg/L.

Constituent	Major Sources	Concentration in Natural Water	Effect Upon Usability of Water
Calcium (Ca^{+2})	Amphiboles, feldspars, gypsum, pyroxenes, aragonite, calcite, dolomite, clay minerals	As much as 600 mg/L in some western streams; brines may contain as much as 75,000 mg/L.	Both calcium and magnesium combine with bicarbonate, carbonate, sulfate and silica to form heat-retarding, pipe-clogging sale in boilers and in other heat-exchange equipment. Calcium and magnesium combine with ions of fatty acid in soaps to form soap suds; the more calcium and magnesium, the more soap required to form suds. A high concentration of magnesium has a laxative effect, especially on new users of the supply.
Magnesium (Mg^{+2})	Amphiboles, olivine, pyroxenes, dolomite, magnesite, clay minerals	As much as several hundred mg/L in some western streams; ocean water contains more than 1,000 mg/L and brines may contain as much as 57,000 mg/L.	
Sodium (Na^+)	Feldspars (albite); clay minerals; evaporites, such as halite (NaCl) and mirabilite ($Na_2SO_4 \cdot H_2O$); industrial wastes	As much as 1000 mg/L in some western streams; about 10,000 mg/L in sea water; about 25,000 mg/L in brines.	More than 50 mg/L sodium and potassium in the presence of suspended matter causes foaming, which accelerates scale formation and corrosion in boilers. Sodium and potassium carbonate in recirulating cooling water can cause deterioration of wood in cooling towers. More than 65 mg/L of sodium can cause problems in ice manufacture.
Potassium (K^+)	Feldspars (orthoclase and microcline), feldspathoids, some micas, clay minerals	Generally less than about 10 mg/L; as much as 100 mg/L in hot springs; as much as 25,000 mg/L in brines.	

Reproduced with permission from F. Van der Leeden, F. L. Troise, and D. K. Todd, *The Water Encyclopedia* (Lewis Publishers, Boca Raton, FL, 1991), pp. 422–423.

Appendix B

Dilutions

Dilutions

You may find that the concentrations of certain ions in your water sample are too high for accurate analysis. If so it will be necessary to dilute the sample to keep it from going out of the range of the analysis. The first step is to pour a small amount of the concentrated sample into a small, clean beaker.

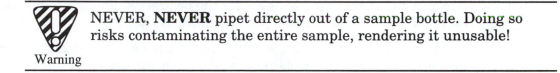

NEVER, **NEVER** pipet directly out of a sample bottle. Doing so risks contaminating the entire sample, rendering it unusable!

Warning

Using a pipet bulb (**NEVER PIPET BY MOUTH!**) pipet a known volume of your **filtered** sample into a **labeled** volumetric flask of the appropriate size and fill the flask to the mark with deionized water so the bottom of the meniscus (the curved line at the water/air interface) just touches the mark.

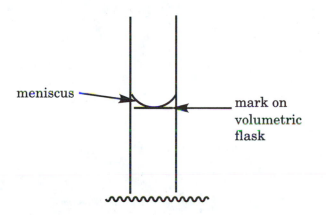

Cover the top of the flask with Parafilm™ or a stopper and invert the flask several times to ensure thorough mixing. You may need to experiment to find the best dilution.

The **dilution factor** can be calculated by dividing the total volume of the diluted sample by the volume of concentrated sample that was used to make up the dilute sample, e.g., if you diluted 5 mL of sample to 100 mL total volume, the dilution factor is $100 \div 5 = 20$.

Here are some tips for perfecting your diluting technique:

- NEVER use a graduated cylinder to do a dilution in which precision is important, always use a volumetric flask and a pipet.

- When pipetting, use the forefinger to control the outflow of liquid, not the thumb. Precise control over the outflow can be had by raising the *back* of the forefinger. Alternatively, use a pipet bulb that permits fine control of dispensing the liquid.

- Dilute EXACTLY to the line, matching the line on the flask to the bottom of the meniscus. Use a Pasteur pipet to add the last few drops. If you overshoot, start over. To eliminate parallax error, the flask should be read at eye level.

- ALWAYS cap the flask with a stopper or with parafilm, invert the volumetric flask, and shake it several times to mix the solution well.

- ALWAYS label the receiving bottle as to the dilution factor.

- ALWAYS rinse the receiving bottle with a small amount of the solution that will be stored in the bottle.

If you are using an autopipettor, here are some special tips:

- Use a clean pipet tip for each sample.

- Be sure pipet tip is on securely. If it leaks, it is not on tight enough.

- Be sure to check that the pipettor is set for the desired volume EACH TIME you pipet. Don't assume it is set for the correct volume.

- Most pipettors have a mechanism for blowing out the last drop of sample from the pipet tip. Be sure you know what that mechanism is for your pipettor and use it.

- Do not submerge the tip too far below the surface of the liquid being pipetted.

- Release the mechanism slowly to avoid splashing the solution up inside the pipettor.

Using Microsoft Excel
for Data Analysis

Update Notice. A new version of Appendix C using Microsoft Excel for Data Analysis (v. 6.0 or 1998) and the accompanying practice data file can be downloaded by visiting this book's website at www.uscibooks.com/kegley2.htm and selecting the link labeled "Appendix C: Download updated version for Microsoft Excel 6.0"

MICROSOFT EXCEL WORKSHEET

Read This Page First!!

Microsoft Excel is a good way to deal with large quantities of numbers in a manageable way. It can carry out a variety of operations accurately, quickly, and painlessly. Below are listed some of the commands within the Microsoft Excel program that you may want to use to manipulate your data. Initially you will be given the file that contains all the chemical data that your lab section collected in this first module. Microsoft Excel will allow you to print out the data and plot relationships in the data, in addition to obtaining the statistics that provide information about how precise your measurements are. The instructions that follow are for versions 5.0 and beyond and do not apply to earlier versions of the program.

How to Find Your Class Data File

Your instructor will either give you a disk copy of your data file or make it available over the World Wide Web. Be sure you have the final version of the class data. Under the **File** menu, go to **Open** to open the file so you can begin work on it. *As soon as you open the data file,* save it *under a different name* by pulling down the **File** menu and clicking on **Save As**. Rename the file and keep the original. If you happen to alter your data file beyond repair, you can always go back to the original and start over.

Warning

If you do something you wish you hadn't, you can **Undo** the action (but only the <u>last</u> action performed) by going to the **Edit** menu and clicking on **Undo**. If the **Undo** command is no longer active, just close the file by pulling down the **File** menu and clicking on **Close**. Respond "No" when it asks if you wish to save the file, then reopen the saved version of the file and start from your last saved version. Obviously, this speaks strongly for saving your file at regular intervals!!

Startup Procedures for the Excel Practice Worksheet

- Start up a computer. If you are using a campus facility, *be sure to bring a 3.5" diskette with you.*

- Click on the Microsoft Excel icon to start the program.

- Proceed to the worksheet on the following page.

EXERCISES IN EXCEL

The following exercises provide practice in carrying out simple plotting and statistical analysis using Microsoft Excel™. Excel has many more capabilities–this is just an introduction to what you will need to analyze the data from the water chemistry experiment.

Exercise 1: Formatting a Data Sheet

Detailed instructions that pertain to this exercise are on page 156.

Format a blank data sheet for the absorbance vs. concentration data you collected on any set of standards used in the colorimetric or AA analysis. Label the columns of data in the topmost row of the column. Format the cells to display the data to the third decimal place (0.000) for the absorbance data column and to the second decimal place (0.00) for the concentration data column. Save the file on your disk by clicking on **Save** from the **File** menu and giving the file a name.

Exercise 2: Entering Data

Detailed instructions that pertain to this exercise are on page 157.

Enter your absorbance and concentration data from one of your analyses into the data sheet you just created.

Exercise 3: Printing the Data

Print the data using the **Print** command from the **File** menu. Printing takes a few minutes. While you are waiting, read on and continue working.

Exercise 4: Creating a Scatter Plot

Detailed instructions that pertain to this exercise are on pages 157-158.

Using the data sheet you just created, make a linear scatter plot (Format 1) of "Absorbance" vs. Concentration." Be sure to create a meaningful title for the plot. Label the *x*- and *y*-axes and do not forget the units on your labels!

Note

Remember that "connect-the-dots" plots **SHOULD NOT BE USED** for the water chemistry data. Use only Format 1 in the Chart Wizard.

Exercise 5: Opening an Existing Data File

Open the file **Anza.practice** by going to the **File** menu and clicking on **Open**. Find the file with the above name and click on it to open.

Exercise 6: Creating a Column Plot

Detailed instructions that pertain to this exercise are on pages 157-158.

Make a **Column** plot (Format 3) of "Site No." vs. "Fecal Coliform." Save your plot as "**FC.Anza**".

Exercise 7: Creating a Stack Column Plot

Detailed instructions that pertain to this exercise are on p. 158.

Make a **Stack Column** plot (Format 3) of "Site No" vs. "Sodium," "Calcium," and "Magnesium." Save the file as "**Cations.Anza**".

Exercise 8: Fitting the Best Straight Line to the Data

Detailed instructions that pertain to this exercise are on page 159.

Go back to the plot of "Absorbance vs. Concentration" you created in Exercise 4 and do a **Linear** curve fit of the data. Note the equation and the R^2 value in your lab notebook. It should be very close to 1.0000. Save the plot.

Exercise 9: Fitting a Second-Order Polynomial to the Data

Detailed instructions that pertain to this exercise are on page 159.

Using the same plot as for Exercise 8, first remove the linear fit by highlighting the line, going to the **Edit** menu, and choosing **Clear**. Inside this submenu choose **Trendline**.

Now return to the plot and choose a **Polynomial** fit, **second order**. A second-order polynomial is the highest you should go for a standard curve. Note the equation and the R value in your lab notebook. Save the plot under a different name than the linear fit.

Which plot gives the best R^2 value, i.e., the value closest to 1.0000? This is your best fit of the data.

Exercise 10: Higher-Order Polynomial Fits

Detailed instructions that pertain to this exercise are on page 159.

Start with a new data sheet and enter the data you obtained for the alkalinity titration. Plot the data using a **Linear Scatter** plot (Format 1), putting pH on the y-axis and amount of acid added on the *x*-axis. Do a **6th-order polynomial** curve fit.

Exercise 11: Calculations with Excel

Detailed instructions that pertain to this exercise are on page 161.

In the file "**Anza.practice**", calculate the total cation concentration (sum of Ca^{+2}, Mg^{+2}, and Na^+) for each site and place the results of the calculations in a new column, naming it "Total Cations."

Calculate the **Average** and **Stdev** for the sodium, calcium, and magnesium data and place this data in two new rows.

Calculate the **Average** for the fecal coliform data. Recalculate it, but this time mask the unusually high points by excluding them from the calculation. Note the difference in the average value.

Exercise 12: Correlation of Water Quality Parameters

Detailed instructions that pertain to this exercise are on ppage 157–158 and page 159.

Using the new column of "Total Cations," make an **x-y (Scatter)** plot of "Total Cations" vs. "Conductivity" to determine if there is a correlation between cation concentration and conductivity.

Under the **Insert** menu, use **Trendline** to fit a **Linear** curve onto the data. Note the R^2 value and the equation of the line. If your R^2 value is over ~0.5, there is some correlation. The higher the R^2 value, the more strongly correlated is your data.

Note When you are looking for a correlation between the concentrations of different ions or between concentrations of ions and another parameter, the **Linear** curve fit is the only one that has any meaning. **Polynomial** fits are NOT appropriate for this type of correlation plot.

Make a note on the plot indicating which data points were excluded from the plot. Detailed instructions that pertain to this exercise are on page 161.

FORMATTING THE DATA SHEET

The data sheet contains a listing of the data in tabular form. The rows and columns can be manipulated in various ways to present the data in the way you choose. The first step in creating a data sheet is to open a new Excel file (**New** under the **File** menu) and format the blank data sheet to include the appropriate number of columns, a title for each column, column width, and data format.

Changes that pertain to significant figures (numbers), alignment, fonts, borders, and patterns can be made from the **Format** menu. Under the **Format** menu, click on **Cells**.... A dialog box will appear. In this dialog box is a series of tabs that you can use, depending on the aspect of the data sheet you wish to change.

Changing the Number of Significant Figures Displayed

To change the number of significant figures that are displayed, go to the **Format** menu and select **Cells**. A dialog box will appear. In this box, select the tab that is called **Number**. In this box there is a list which appears in the left side. Choose **Number** in this list. In the **Code Box** type in how many decimal places you wish to be displayed. For example, if you want 2 places past the decimal point displayed, you would type in 0.00 in the code box. Excel automatically rounds the number entered to fit the chosen format.

⚠️ Before you start entering data in a new spreadsheet, format the cells for the correct number of significant figures.

Note

Inserting or Deleting Rows and Columns

Excel defines the **first row** (horizontal) of the spreadsheet as the names of the columns. Thus, if you name a column, this will eventually be the name of the data series in the column that will be plotted on a graph of the data. If you do not assign names to the columns, the letters A, B, C, ..., will be used as the default. Excel defines the **first column** (vertical) of the spreadsheet as the names of the rows. These names then designate the points along the x-axis. If you do not assign names to the rows, the numbers 1,2,3,..., will be the default.

To insert a column, first highlight the column to the right of where you wish the new column to be. Go to the **Insert** menu and click on **Column**. If you wish to insert a row, first highlight the row below where you wish the new row to be. Go to the **Insert** menu and click on **Row**. To delete a column, select the whole column by using the mouse to click on the letter that is directly above the appropriate column. For example, if the column you wish to delete is under the letter C, you can select the whole column by clicking on C. Once this is done, go to the **Edit** menu. Under the **Edit** menu choose **Delete**.

To delete a row, the procedure is similar except that instead of choosing the appropriate letter, you choose the appropriate number on the far left of the spreadsheet. This selects the whole row that you want to delete. After selecting a row, go to the **Edit** menu and select **Delete**.

ENTERING THE DATA

Data can be entered in two ways, either by direct data entry (i.e., typing the numbers in individually) or by importing the data from another file. Moving between cells can be accomplished by using:

- Tab key

- Arrow keys

- Mouse

- Return key

Moving Rows and Columns

Useful commands for moving data within a file or from one file to another are found under the **Edit** menu.

- **Cut:** Removes highlighted data from the data sheet to the clipboard.

- **Copy:** Copies the highlighted data to the clipboard, leaving the data sheet intact.

- **Paste:** Pastes the contents of the clipboard into a data sheet.

An alternative (faster) method for moving rows and columns is to highlight the row or column you wish to move. If you then click on the highlighted cells, you can drag them to a new location. Be careful! If you drag the highlighted cells on top of occupied cells, the data in the occupied cells will be overwritten. Holding the **shift** key down while dragging the cells allows you to insert the highlighted cells instead of overwriting.

PLOTTING THE DATA

Once the data have been entered, it is easy to create a graph of the data. Choose **Chart Wizard** from the myriad of **Buttons** at the top of the page under the menu selections. **Chart Wizard** is the button that shows an *x-y* axis with a bar graph.

Also incorporated in this picture is a wand (denotes the wizard part.) Do not be confused with the button that shows a square, a triangle, and a circle. This is a drawing utility. Once the **Chart Wizard** is chosen, a series of dialog boxes appear to help guide you through the graphing process.

Choosing the Data to Plot

Click on the **Chart Wizard** to begin the process. Your cursor will change to a cross hair. Position the cross hair somewhere on the spreadsheet (not on top of your data) and hold the mouse button down. With the mouse button still depressed, drag the cross hair down and to the right so a box appears. Inside the box is where your graph will be. You can make the box any size. After you let go of the mouse button, a dialog box will appear asking you for the range of the plot. Excel uses the term "range" to define the columns or rows of data in the spreadsheet. You will need to tell the computer what range of rows and columns are to be plotted. There are three ways to enter the desired range.

- **Method 1: Data in Adjacent Columns**

  If the data you wish to plot is entered in adjacent columns, simply highlight all of the data you wish to plot to define the range. You may also do this **before** you click on the **Chart Wizard** button.

- **Method 2: Data in Columns NOT Adjacent to Each Other**

  If the data you wish to plot is *not* in adjacent columns, begin by highlighting the first column of data. This results in an entry into the "Range" box that defines this column of data. To enter the next column of interest, put a comma after the first range, then go back to the data sheet and highlight the next column. This method is awkward and slow. It is usually easier to move the data columns (see page 157) to make them adjacent so you can use Method 1.

- **Method 3: Manual Entry of the Range**

  You may also define the range manually. If you choose to do this, the syntax is always "=$(letter of first column)$(number of first row):$(letter of last column)$(number of last row)", for each column of data. Separate columns of data by a comma.

Choosing the Type of Plot

The next two dialog boxes will ask you what type of graph you wish to plot. There are two types of plots you will find most useful for water quality data, scatter plots and column plots.

X-Y (Scatter) plots

When you do an *x-y* scatterplot, the usual intent is to see if there is a noticeable correlation between the *x*-values and the *y*-values. For example, perhaps you would like to know if the amount of chloride in your samples is correlated to the amount of sodium in the samples. If you have a positive correlation, you could then postulate that the source of the sodium and chloride in the sample is sodium chloride, and not sodium sulfate and calcium chloride. The most useful type of **x-y (Scatter) Plot** that Excel offers is **Format 1**.

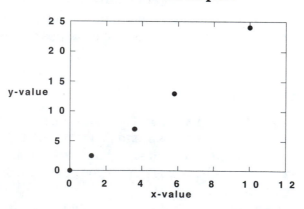

X-Y Scatterplot

Column plots

A column is a vertical bar graph with the value of the variable plotted on the *y*-axis and the descriptor of the value on the *x*-axis. This type of plot is good for depicting the value of a water quality parameter (*y*-axis) as a function of site (*x*-axis). One specific type of column is called a **stack column**. This is a vertical bar graph with the values of several variables plotted one on top of the other on the *y*-axis and the descriptor of the variables on the *x* axis. The most useful type of **Column** plot that you can choose from the Chart Wizard is **Format 3**.

Column Plot

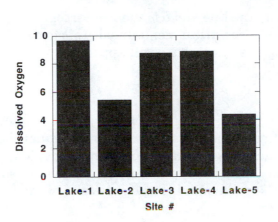

Defining Which Data Goes on the *x*-Axis

The fourth dialog box asks whether the data series is in columns or rows. This box also asks which column to use as the *x*-axis labels. If you enter 0, the generic labels of 1,2,3,..., will be used along the *x*-axis. If you wished to plot column A vs. column B, it is necessary to specify the number "1" in this box so the first column (column A) is used as the descriptor for the values on the *x*-axis.

Axis and Chart Titles

The axis and the chart titles can be modified in the final dialog box of the Chart Wizard. Type in a descriptive title for your plot and be sure to label your axes, including units where applicable.

CURVE FITTING

Excel offers a variety of curve-fitting routines that allow you to fit the data to a function that allows prediction of unknown values. The following paragraphs discuss the types of curve fits most useful for water analysis and some potential applications of each.

Setting Up for Curve Fitting

To begin the process of curve fitting, you must first double click on the graph so that it has a fuzzy border. Next, click on any one of the data points to highlight the data series on the graph. Go to the **Insert** menu and choose **Trendline**. A dialog box will appear. Choose the **Options** tab and **X** the boxes to show (1) the equation describing the line and (2) the R^2 values. Choose the **Type** tab, then select one of the following types of fitted curves.

Linear Fits

The **Linear** curve fit uses linear regression analysis to determine the best straight line for a series of data points. It is useful in fitting a standard curve for a quantitative analytical procedure or depicting a linear relationship between two parameters. The equation is in the general form of

$$Y = mX + b$$

where m is the slope of the line and b is the Y-intercept. To insert a linear curve fit, go to the **Type** folder inside the **Trendline** menu. Inside this folder choose **Linear**, and a best fit line based on a linear regression analysis will appear.

Polynomial Fits

A **Polynomial** curve fit calculates the coefficients of a polynomial that describes the data. It is useful in fitting a nonlinear standard curve for a quantitative analytical procedure (i.e., AA analysis for sodium, magnesium, or calcium) or depicting a relationship between two parameters. The more complex and nonlinear the relationship is between the data, the higher the order of the polynomial. The equation is of the form

$$Y = c_0 + c_1 X + c_2 X^2 + c_3 X^3 + ... + c_n X^n.$$

To insert a polynomial fit onto your data, you need to click on **Polynomial** in the **Type** folder. After **Polynomial** has been selected, the **Order** box will be highlighted. You must then specify which order you want. Higher-order polynomials are used to fit lines with significant curvature. It is useful to vary the order to determine which gives the best fit to the data.

MAKING THE PLOT LOOK LIKE YOU WANT IT TO

To make changes to the formatting of a plot, you will first need to highlight it. This is done by double clicking on the graph. As you will see, the plot is now surrounded by a thick fuzzy border.

Changing Plot Size

To change the size of the plot, place the pointer at the lower right corner of the graph. Depress the mouse button, and while keeping it depressed, drag the mouse until the desired size and shape are obtained.

Changing the Looks of the Axes

Once the graph is highlighted, double click on the axis that you want to modify. This activates a dialog box that enables you to modify the scale and tick marks shown on the axis, the font on the axis label, the text alignment, and the way the numbers are shown. One potential problem is that Excel may not have room to place all of the labels on a particular axis. The program then omits some of the labels. This can be a problem if the labels are, for example, site numbers. Unless your site numbers are always labeled consecutively, you may not know which data point corresponds to which site under these circumstances. There are two possible solutions to this problem. You may need to use both of these tricks to get the plot to look right.

- Stretch the graph out sideways using the grab box in the lower right-hand corner, so that the x-axis takes up more space and there is now more room for the labels.

- Change the font size to something smaller, down to 6-point size.

Inserting Comments on a Plot

Often it is useful to insert a comment on your plot. For example, you may have a site where no data were collected and would like to let your reader know that the concentration is not zero, but there is no data. You can easily add comments to any Excel plot.

To do this, you must first activate the plot by double clicking on it. Type in the word or phrase you wish to add. It will likely show up on the plot in a random location. When you are finished typing, you can select the text box and move it to the desired location. You can also change the format of the text by highlighting the textbox and going to the **Format** menu and choosing **Selected Object**. There are a variety of options within this menu for changing font, alignment, color, etc.

CALCULATIONS

Excel allows you to carry out mathematical calculations on entire columns of data. An example of this is if you have data in columns A and B that you want to add together and place in column C. To add cell A1 to cell B1, in cell C1 you would type an equal sign, "=", and then click on cells A1 and B1 individually. What you should see appear in cell C1 is =A1+B1. Press enter, and the computer completes the calculation. You can also manually type "=A1+B1".

Now consider the situation where there are 10 rows of data and you would like every entry in column C to be the sum of columns A and B. It would require some work to go through and individually write out the equation for each cell. Here's a shortcut.

- Generate the first equation on either the top or bottom row, as instructed above.

- Now highlight the remainder of the cells in column C that will contain the same formula, *including the one in which you originally typed the equation*. Proceed to the **Edit** menu and choose **Fill**. If all the highlighted cells are below the equation, then choose **Down**. If the highlighted cells are above your formula, then choose **Up**. You may also **Fill Left** or **Right**.

The following is a list of arguments that can be used in mathematical equations.

- + Add
- - Subtract
- * Multiply
- / Divide
- ^ Power (exponent)

STATISTICS: THE EASY WAY

Statistical operations are similar to the calculations shown above in that they are long equations. Excel already contains the equations defining a variety of statistical calculations.

To perform a statistical operation, go to **Function** under the **Insert** menu. A dialog box will appear. Choose **Statistical** in the **Function Category** box. Statistical functions will appear in the **Function Name** box that is adjacent to the **Function Category** box. Choose your function. The ones you will be using the most are **Average** and **Stdev** (standard deviation).

The next dialog box asks for the data for which statistical information is desired. Highlight the cells you wish to analyze. If your cells are not contiguous, use a comma to separate the range entered. After you are finished, the equation and range will appear in the cell you specified that it would be in. Press enter and the program performs the calculations.

Errors in entering data or formulas are always possible. Check your result to see if it makes sense!

Note

Masking Bad Data Points

You may wish to carry out statistical calculations on a data set, but omit bad data points. Consider the situation if you wanted the average of A1 to A20, but wanted to mask cells A4 and A18 since they are very different from the other values and you suspect error. There are several ways to accomplish this:

Method 1

To begin, insert the function **Average** in a cell at the bottom of the column. When the dialog box appears to asks for the numbers to average, highlight A1 through A3. Next put a comma after A3 and then highlight A5 through A17. Then put a comma after A17 and highlight A19 through A20.

Method 2

An alternate way to mask bad data points is to just remove the entire row of data from its location in the middle of the data set to the bottom of the spread sheet. Then you may simply highlight a contiguous set of cells.

SHUTDOWN

1   Save your files.

2   Quit the Excel program by going to the **File** menu and clicking on **Quit**.

Index